BOTANICAL RESEARCH AND PRACTICES

NATIVE SPECIES: IDENTIFICATION, CONSERVATION AND RESTORATION

BOTANICAL RESEARCH AND PRACTICES

Additional books in this series can be found on Nova's website under the Series tab.

Additional E-books in this series can be found on Nova's website under the E-book tab.

BOTANICAL RESEARCH AND PRACTICES

NATIVE SPECIES: IDENTIFICATION, CONSERVATION AND RESTORATION

LLUVIA MARÍN
AND
DIMOS KOVAČ
EDITORS

Nova Science Publishers, Inc.
New York

For permission to use material from this book please contact us:
Telephone 631-231-7269; Fax 631-231-8175
Web Site: http://www.novapublishers.com

LIBRARY OF CONGRESS CATALOGING-IN-PUBLICATION DATA

Native species : identification, conservation, and restoration / editors, Lluvia Marmn and Dimos Kovac.
p. cm.
Includes bibliographical references and index.
ISBN 978-1-61470-613-7 (hardcover)
1. Endemic plants--Identification. 2. Endemic animals--Identification. 3. Endemic plants--Conservation. 4. Endemic animals--Conservation. I. Marmn, Lluvia. II. Kovac, Dimos.
QK101.N28 2011
333.95--dc23

2011024361

Published by Nova Science Publishers, Inc. † New York

CONTENTS

PREFACE

This book presents current research from across the globe in the study of the identification, conservation, and restoration of native species. Topics discussed include soil microbiology influencing biodiversity conservation of natural ecosystems; the native breed of domestic cattle in Southwest England; native plants of northeastern Mexico and their potential in the restoration of damaged ecosystems; Berberis buxifolia, a native fruit species of Patagonia; the contribution of the native earthworm species of Assam state in northeast India and a review of the native species of salt steppes in Algeria.

Chapter 1 - Since for terrestrial ecosystems, land-use change probably will have the largest effect on biodiversity, research on native species of both plant and fungi have received more attention for restoration of degraded ecosystems and for mitigation of the expected effects on biodiversity. Very few studies in South America have pointed to the importance of the soil microbiological and biological processes influence upon biodiversity conservation of natural ecosystems. Research on the mycotrophy of plants is thus fundamental to the understanding of the ecophysiological characteristics and therefore for improve their conservation. The same applies for endangered vegetation, especially under extreme climatic and edaphicenvironmentalconditions. Until recently, the role of native plant endophyte (and, also of symbiotic bacteria or fungi) had received little attention. Studies of aquatic plants, types of endophyte and dominant species colonizing plants offer new reports and point to the urgent need for further research. In paricular, reports of dark septate endophytes in roots of plants have received increasingly more attention as possible mutualistic. However, it must be kept in mind that symbiosis type is also influenced by environmental factors. Little is known about the soil biota and arbuscular mycorrhizal fungi present in roots of most plant species also dominating in tropical dry forests and in the Andean region, which could help in dealing with the uncertainties in future climate and vegetation. Very few studies have been devoted to the consequences of the plant symbioses on the ecosystem function. The aim of this review is to explore currently available information on soil microbiology influencing biodiversity conservation of natural ecosystemsand to speculate about the management of natural sitesfocusing on root symbioses. Case studies of the Brazilian dry forest/woodlandand the Andean Puna are mentioned.

Chapter 2 - Early British explorers and settlers introduced their domestic livestock to new lands. Many species and breeds were indigenous to the place of origin of the introducers. The Devon breed of cattle is one example of such introductions. Cattle recognizable as the progenitors of the Devon type appeared in its native county in southwest England in Domesday times at the end of the eleventh century and a beast of Devon colour was given as heriot to Tavistock Abbey in the mid-fourteenth century (Finberg, 1951). There is no direct evidence of a Devon "breed" before the eighteenth century but a type was beginning to be selected and fixed by local stockmen from this period on. A Herd Book (pedigree records) was established in 1851 and a Breed Society formed in 1884. Early in the twentieth century the Devon was the second most numerous breed in the UK. Numbers declined after the Second World War in particular with the influx of "continental" breeds that were introduced in response to changing customer demand for meat but have since recovered to some extent. Red cattle from Devon were taken to the USA by pilgrims early in the seventeenth century and to Australia some two hundred years later. Devons, as such, have been exported to about 30 countries with numbers varying from as few as one to more than 800. In some countries they were used for crossbreeding and especially to produce beef from dairy herds. In others, notably Australia, Brazil, New Zealand, the USA and until recently South Africa they have been maintained as pure breeding herds with their own Herd Books and Breed Societies. In the early twentyfirst century the Devon, bred for thriftiness, fertility, docility and easy management is ideal as a beast that provides quality meat to a hungry world concerned about economic and environmental costs.

Chapter 3 - Native plants naturally occur in the region in which they evolved and adapted to the physical and biotic processes characteristic of that region, e.g. climate, soil, time of rainfall, drought, and frost as well as interactions with other species inhabiting the local community. Thus native plants possess certain traits that make them uniquely adapted to local conditions providing a practical and ecologically valuable alternative for landscaping, conservation and restoration of degraded ecosystems. Mexico's ecosystem is a vast one as the country covers a wide variety of terrains, soils and climates with different geography characteristics which favor the development of native species. For instance, in those arid and semiarid regions of northeastern Mexico there is an extensive native plant list of more than 3,100 species. Regions of the coastal plain of the Gulf of Mexico are dominated by a dense diversity of native plants. The vascular flora comprises 159 families, 1034 genera, 3181 species and 109 intraspecific taxa. The most diverse families are: Asteraceae (148 genera and 481 species), Poaceae (97 genera and 301 species), Fabaceae (76 genera and 258 species), Cactaceae (37 genera and 139 species) and Euphorbiaceae (15 genera and 85 species) among others. Native plant inventory in northeastern Mexico is mostly composed of deciduous and evergreen or perennial species which are characterized by a wide range of growth patterns, diversity in leaf longevity, dynamics of growth and contrasting phenological developments. This resource is used by the inhabitants of the region in various ways such as forage for livestock, coal, timber, shelving, food, herbal, medicine and seed supply for reforestation and ecological restoration. Additionally, the native plant provides a habitat for wildlife and a cover layer that prevent soil erosion. For these reasons native plants develop various morphological and physiological characteristics suitable for adaptation to adverse environmental factors, particularly drought stress and high temperatures. Numerous vegetal species account for the bulk of the bush vegetation providing a characteristic landscape. The most important of these plants consist of various species of *Acacia* including *A. rigidula*, *A. farnesiana*, *A. berlandierii*, *A. smallii* and *A. tortuosa*, and some other important species such as *Prosopis laevigata*, *P. glandulosa*, *Leucaena leucocephala*, *Celtis pallida* (desert hackberry), *Condalia ericoides* (javelina bush), *Leucophyllum frutescens* (cenizo or sometimes called purple sage), and *Aloysia wrightii* (common bee-brush or white brush). These species favor the retention of soil due to their high sprouting potential, ease of regeneration after frost and forest fires in addition to being accepted by the people so they may be used successfully to restore degraded areas and initiate programs aimed to the protection and conservation of forest elements in northeastern Mexico.

Chapter 4 - Recent findings observed on *Berberis buxifolia* plants grown from different spontaneous populations on Tierra del Fuego island (Argentina) are reported, as related to: (I) the floral morphology and evolution of the floral phenological phases; (II) the fruiting efficiency; (III) the pattern of fruit growth and evolution of some of the chemical attributes; and (IV) the variability found in the leaves and seeds. Flowers are solitary, axillary, bisexual, actinomorphic and hypogynous. Petals present the nectaries characteristics of the genus. Fruit is a black-blue berry (8-10 mm diameter) with an average of 3-18, moon-like shaped seeds. Bloom occurs at different times, and its beginning and duration is an accession characteristic, depending also on ecological factors. Nevertheless, seven floral phenological phases were recognized. Phenophases of blooming were measured during the spring of 2010. The length of blooming of the *B. buxifolia* accessions studied was 15 days between start of blooming and petal fall. Fruit number per shoot varied with the orientation and during the fruiting, being higher in the north (4.3) and west (4.0) orientations than in those from the east (2.0) and south (1.3), attaining maxima of 8.8 and 7.8 fruits per shoot respectively at day 38. Fruit efficiency then began to decrease during the ripening, attaining values of 20 to 25% for those shoots from the north and west orientations, and values of 13 to 15% for the shoots from the east and south. Fruit growth showed a typical double sigmoid curve, and it was related to compositional changes. By day 126 from full flowering phase, soluble solids (30.6 °Brix) and anthocyanin content (752.7 mg/100 g fruit fresh weight) were near their maximum values, while the total titratable acidity was at a minimum value (2.2 %). The present investigation confirmed the previous description of leaves; nevertheless a considerable variability has been observed in populations of *B. buxifolia*. Leaf size ranged from 0.04 to 2.08 cm^2 (average 0.5 cm^2); similar results were found for leaf shape (major axis/minor axis of leaf blade ranged from 1.26 to 6.25); furthermore, leaves color analysed with the RGB scale varied among populations and plants. In *B. buxifolia* seed number per fruit may range from 1 to 15, with an average 7 seeds/berry; results indicated that clonal selection may be applied for this trait. Seed area ranges from 5 to 16 mm^2, with an average of 9 mm^2; seed are lunular, mostly elongated and very elongated in shape.

Chapter 5 - Assam is a high earthworm diversity region and constitutes 26.2% of the total known Indian species; among them 22% are native of India. The earthworms have innate capability to convert various industrial wastes into nutrient rich vermicompost with higher agronomic value besides enhancing soil fertility. A survey of earthworm bioresources with particular reference to native species was, therefore, undertaken during 2010 in Guwahati, Goalpara and Bangaigaon districts of Assam, hitherto, uninvestigated in this direction. The survey provided a total of six species of earthworms, namely, *Lampitomauritii*Mauritii, *Perionyxexcavatus* Perrier, *Metaphireanomola*Michaelsen, *Drawidanepalensis*Michaelsen, *Ocnerodrilusoccidentalis*Eisen and *Metaphireposthuma*Vallant; of these later two species are exotics. The study indicates that *Lampitomauritii*Mauritii and *Metaphireposthuma*Vallant are the predominant species for the area investigated. Need for such investigations with a view to target native species of earthworms and conserve their biodiversity in underexplored or unexplored areas of India and abroad has been re-stressed.

Chapter 6 - *Atriplex halimus* subsp. *schweinfurthii* (Chenopodiaceae) is perennial spontaneous species of Algerian salt steppes with an excellent tolerance to drought and salinity. This species is very palatable and is a preferred fodder shrub of livestock during dearth periods. Endowed with a complex root system and a considerable biomass, *A. halimus* is an efficient and a relatively cheap barrier against erosion and desertification. Its ligneous wood is interesting source of energy. In view of this, *A. halimus* plays a major role in the rehabilitation of the degraded lands, especially in the Mediterranean arid and semi-arid areas. Salt-affected soils will remain unproductive unless efforts are made to rehabilitate them with highly salt-tolerant plants (halophytes). It goes without saying that salty areas such as salt steppes of Algeria can benefit from being re-vegetated by planting saltbush and farmers can also obtain fodder for their livestock from this halophytic shrub. This paper is to review the importance of *A. halimus* as forage crop for possible cultivation in saline steppes of Algeria.

In: Native Species
Editors: L. Marin and D. Kovač
ISBN: 978-1-61470-613-7

Chapter 1

NATIVE SPECIES FOR RESTORATION AND CONSERVATION OF BIODIVERSITY IN SOUTH AMERICA

Marcela C. Pagano,[1] Mónica A. Lugo,[2] Francisca S. Araújo,[1] Marcela A. Ferrero,[3] Eugenia Menoyo,[2] and Diego Steinaker[4]

[1]Biology Department, Federal University of Ceará, Fortaleza, Ceará, Brazil
[2] IMIBIO-CONICET-UNSL- Facultad de Química, Bioquímica y Farmacia, San Luis, Argentina
[3] PROIMI-CONICET-UNT-San Miguel de Tucumán, Tucumán, Argentina
[4]INTASan Luis,Villa Mercedes, San Luis, Argentina

ABSTRACT

Since for terrestrial ecosystems, land-use change probably will have the largest effect on biodiversity, research on native species of both plant and fungi have received more attention for restoration of degraded ecosystems and for mitigation of the expected effects on biodiversity. Very few studies in South America have pointed to the importance of the soil microbiological and biological processes influence upon biodiversity conservation of natural ecosystems. Research on the mycotrophy of plants is thus fundamental to the understanding of the ecophysiological characteristics and therefore for improve their conservation. The same applies for endangered vegetation, especially under extreme climatic and edaphicenvironmentalconditions. Until recently, the role of native plant endophyte (and, also of symbiotic bacteria or fungi) had received little attention. Studies of aquatic plants, types of endophyte and dominant species colonizing plants offer new reports and point to the urgent need for further research. In paricular, reports of dark septate endophytes in roots of plants have received increasingly more attention as possible mutualistic. However, it must be kept in mind that symbiosis type is also influenced by environmental factors. Little is known about the soil biota and arbuscular mycorrhizal fungi present in roots of most plant species also dominating in tropical dry forests and in the Andean region, which could help in dealing with the uncertainties in future climate and vegetation. Very few studies have been devoted to the consequences of the plant symbioses on the ecosystem function. The aim of this review is to explore currently available information on soil microbiology influencing biodiversity conservation of natural ecosystemsand to speculate about the management of natural sitesfocusing on root symbioses. Case studies of the Brazilian dry forest/woodlandand the Andean Puna are mentioned.

Keywords: Native vegetation types, endangered species, soil microbiology, native mycorrhiza, South America

INTRODUCTION

Arid and semiarid areas are spread throughout South America, making up different regions, subregions and provinces of this continent biogeography. Brazil and Argentina are vast countries, their extensive surfaces including various biogeographic areas and harboring the largest human populations in South America. In view of this, the conservation *status* of their natural regions has been changing drastically in the last years. Given that for terrestrial ecosystems land-use change probably will have the largest effect in biodiversity, altering ecosystem processes and changing the resilience of ecosystems (Chapin et al. 2000), research on land native species of both plant and fungi have been receiving increasing attention. During the last years, the agronomic practices have been changing to provide the sustainability of those systems, as well as the sustainable use of natural resources, which allowed to the increasing need to study biodiversity and its conservation.

As Palm and Chapela (1997) state, "human populations depend upon a vast number of other species (e.g., fungi and their benefits to humanity), which means that the quality of our life is related directly to health of the global ecological web" (Systematic Agenda 2000 1994). All over the planet, food chains and webs have decomposers (Fungi and Bacteria) as basic pieces of the soils puzzle, and Fungi are also main nutrient recyclers due to their amount of biomass in soil. For example, most useful topics studies for the success of sustainable development (as conservation of biodiversity and management among others) included Fungi that paradoxically is one of the most under-exploited, under-studied and understood biological group (Palm and Chapela 1997).

It is known that Fungi and Bacteria play a significant role in global C and N cycling and are responsible for about 90% of all organic matter decomposition (Swift et al. 1979). Soil microbes also play an integral role as determinants of the diversity–productivity relationship (Schnitzer et al. 2011). Moreover, soil microbial communities are many times more diverse than their associated plant and animal communities (Whitman et al. 1998, Hawksworth 2001), whereas fewer than 5% of all microbes have been identified and described (Schleifer 2004). While it is broadly recognized that microbes are critical for driving soil nutrient cycling, the role of microbial community structure in determining ecosystem process rates is not well characterized (McGuire et al. 2010). Bacterial populations play an important role in carbon and nitrogen cycling in deserts systems. In nutrient-limited soil, plant demand may exceed soil sources; bacteria are thus essential to nutrient translocation from the soil to the roots. In addition, bacterial nutrient movements are coupled with moisture content in soil, and soil water content is a crucial factor in arid regions (Dhillon and Zak 1993).

It is also known that plant diversity is beneficial for the provision of ecosystem services as the aboveground plant biomass gives useful plant products such as food, fodder, timber and firewood (Díaz et al. 2006, Schmid et al. 2009, Quijas et al. 2010). Moreover, plants have positive effects on erosion control, through increased belowground biomass contributing to greater cohesion of soil (Gyssels et al. 2005) as well as soil fertility regulation by arbuscular mycorrhizas (AM) as providers of nutrient supply from soil (Quijas et al. 2010). It is also known that resistance to plant invasions can consistently increase with plant diversity (Díaz et al. 2006, Schnitzler et al. 2007).

Given current knowledge, mycorrhizae constitute an important component of the ecological process to be applied for revegetation goals (include restoration, reclamation and rehabilitation), and along all of these objectives, mycorrhizal beneficial effects, inoculation methods and management are shared by these different revegetation purposes (Allen et al. 1997). Rare or threatened species also have higher conservation value because they add to national biodiversity (Duelli and Obrist 2003) and there are red list species (for plants and animals), which warn humans about the loss of species. Unfortunately, little has been pointed out about rare species of soil microorganisms.

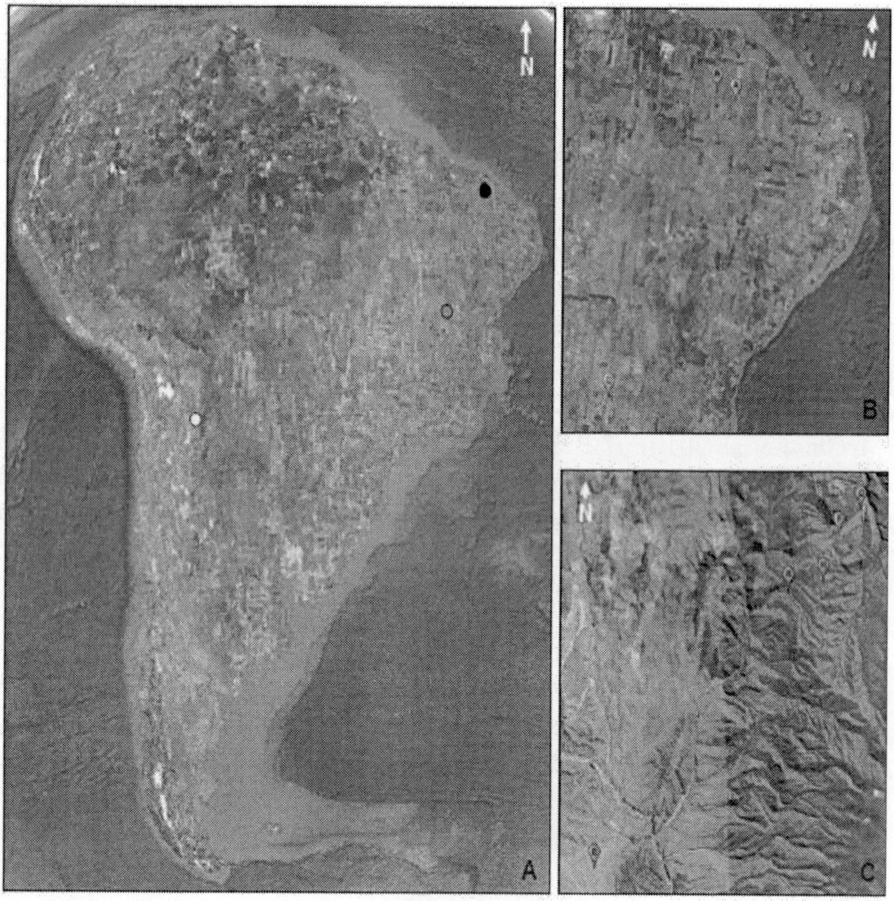

Figure 1. (A) Location of studied sites in Argentina, white circle, and Brazil, grey and black circle, at 5579 miles. (B) Distribution of study sites in Brazil at 1644 miles: A and B, corresponding to Serra das Almas; C, corresponding to dry forest of Minas Gerais; (C) Distribution of study sites in Argentina at 24 miles: the different letters corresponding to different sites in the Puna. Images obtained in Google earth location (Google earth 6.0.2 2010, accessed April 29th, 2011).

The aim of the present chapter is to provide an outline of our present knowledge of native species that are related to restoration and conservation of biodiversity in South America. Both plant and fungi, especially arbuscular mycorrhizal fungi (AMF) and dark septate endophytes in roots of plants (DSE), are involved and their natural occurrence and responses to several environmental stresses and impacts need to be studied in order to point out their benefits as ecosystem services. Soil microbial relationships with native hosts and their conservation status in South America, focusing in particular on arid and semiarid regions of Argentina and Brazil (Figure 1), will be discussed.

Potential Importance of Biodiversity for Ecosystem Restoration

"Ecological restoration is the process of assisting the recovery of an ecosystem that has been degraded, damaged, or destroyed" (Bradshaw 2002). International societies such as SER - Society of Ecological Restoration -, promote ecological restoration as a means of sustaining the diversity of life on Earth, envisaging the re-establishment of ecologically healthy relationships between nature and culture. The main object of restoration is the self-supporting of ecosystems, providing resilience to disturbances without extra-system assistance. Thus, ecosystem sustainability is an important restoration outcome. The SER Restoration Primer (2004) proposed nine ecosystem features as a guideline for measuring restoration success, thus expanding the three features widely applied in practice (diversity, vegetation structure, and ecological processes) (Ruiz-Jaen and Aide 2005). Restored ecosystems having the following features: (1) similar diversity and community structure in comparison with reference sites; (2) presence of native species; (3) presence of functional groups necessary for long-term stability; (4) capacity of the physical environment to sustain reproducing populations; (5) normal functioning; (6) integration with the landscape; (7) elimination of potential threats; (8) resilience to natural disturbances; and (9) selfsustainability, was suggested (Ruiz-Jaen and Aide 2005).

Ecological processes such as nutrient cycling and biological interactions (e.g., mycorrhizas or herbivory) are important because they give information on the resilience and the long-term functioning of the restored ecosystems. For example, over 468 articles analyzed by Ruiz-Jaen and Aide (2005), plant and arthropod richness were common measures of diversity recovery in the studies on restoration, and those studies mainly considered only one group of organisms. In vegetation structure analysis, the most common variables measured are plant cover, density, biomass, and height. Aboveground biomass of plants provides products (food, fodder, timber, firewood) as an ecosystem service, and their belowground biomass provides erosion control (Quijas et al. 2010). Furthermore, in conservation projects, vertebrate are more frequently studied than plants or invertebrates, mammals having the largest proportion of the global number of species currently threatened with extinction (Chapin et al. 2000). For example, some species such as*Cryptotis nelsoni* (Nelson's Small Eared Shrew) are thought to be extinct by habitat loss due to logging, cattle grazing, induced fires and crops in Veracruz, Mexico (IUCN 2010).

Among ecological processes in restoration studies, biological interactions measures are the most common; nutrient pools hold a second position, followed by soil organic matter content. Publications on restoration have been mostly done in North America (mainly United States), Australia, and Europe (Ruiz-Jaen and Aide 2005 and references therein).

With regard to ecological processes, Bacteria are one of the most important groups of soil microorganisms when nutrient cycling is considered. However, few reports show the quantification of environment services by bacteria (May 2011, Maron et al. 2011). It is noteworthythat ecosystem services are provided by biodiversity in an intricate manner and that there is no possible substitute for these services without environmental problems such as those caused by advances in technologies made in recent decades. Recently, Berg and Smalla (2009) stressed the importance of the rhizosphere in terrestrial ecosystems and its participation in ecosystems services (plant nutrition, health and quality, microorganism-driven carbon sequestration, nutrient cycling). Other recent studies pointed the influence of the two factors, plant species (Pagano et al. 2011) and soil type (Oehl et al. 2010), on rhizosphere-associated microbial communities.

It has been previously shown that arbuscular mycorrhizas (AM) have a critical influence on the regrowth of plants in disturbed areas (eg. Moorman and Reeves 1979, Reeves et al. 1979, Miller and Jastrow 1992, Torpy et al. 1999, Requena et al. 2001). However, most studies were conducted in ecosystems from the Northern Hemisphere and Australia.

First studies by Herrera et al. (1993) in Spain showed that appropriate revegetation strategies (inoculation with selected rhizobia and AM) can improve plant survival and biomass. The use of native species was stressed and reclamation (use of species with similar ecological functions as for restoration) was recommended after a 4-year trial due to the exclusive establishment of the native shrub legumes. More recent reports, such as Baar's (2010), pointed out the benefits of applying AMF for plant restoration projects in The Netherlands, encouraging the application of mix inocula of these fungi. They showed the occurrence of AMF and colonization in native field plants and through the use of pot experiments (two commercial mixtures containing different AMF), a higher aboveground plant biomass when mix inocula was applied.

In South America, pioneer studies in La Gran Sabana (Cuenca et al. 1998), Venezuela, showed a low diversity of AMF in disturbed or revegetated (with *Brachiaria decumbens*) sites in spite of a higher spore density. Later, Lovera and Cuenca (2007) stressed the vulnerability of AM diversity to disturbances, especially for some *Scutellospora* and *Gigaspora* (spores of AM of larger size).

Pioneer research by Siqueira et al. (1989) in the southeast of Brazil also showed the vulnerability of some AM species to land change. Also in Brazil, AM interactions with early successional woody species were considered as the main tool to be used in revegetation projects (Zangaro et al. 2003). Studies by Pagano et al. (2009, 2010, 2011) on the AM communities in the Brazilian dry forest/woodlandand in riparian forest restoration experiments (Pagano and Cabello 2011) confirmed the occurrence and potential benefits of these fungi. Thus, if roots remaining after disturbance will help to accelerate sucession e.g. in tropical dry sites (reviewed by Quesada et al. 2009), then highly mycorrhizal-dependent tree hosts could maintain higher AM propagules in soil, which will take less time to colonize new roots.

In Argentina, studies were carried out with restoration purposes (Cabello 2002 and references therein, Becerra et al. 2010). Dr Marta Cabello's pioneer work was conducted in hydrocarbons polluted areas, where data regarding native AMF, its isolation, quantification of infectivity of indigenous propagules and its effectiveness could be applied in bioremediation programs. Her results confirmed the beneficial effects of AMF used as inoculants due to the increased host's efficient adaptation to contamination. Furthermore, those results suggested a huge importance of the selection of AMF species to be included in revegetation practices.

Becerra et al. (2010) studied AMF infectivity in Central Argentina populations of Polylepis australis, a South Americannear threatened species, included in the category 3 of the Argentinian list of threatened native species (http://www.lista-planear.org). Although this AMF-DSE symbiotic species (Menoyo et al. 2007) is still not considered as a vulnerable species (http://www.ambiente.gov.ar), P. australis is highly prone to overgrazing and anthropogenic perturbation, and this negative situation is getting worse due to the species' restricted growing area in mountain ecosystems. Moreover, environmental disturbances by livestock were reported to negatively affect P. australis mycorrhizal colonization and positively impact AMF spores's density (Menoyo et al.2009). The Puna ecosystem has also been analyzed through a consideration of basic ecological information such as AM fungal diversity, AM root colonization, rhizospheric and soil bacteria. These data could certainly be applied to propose viable sustainable agricultural practices oriented towards restoration and conservation of this natural Andean system. Ultimately, understanding the interactions of AMF and the dark septate endophytes (DSE) with plants will help us tounderstand community structure and function and how these will respond to environmental change.

Biodiversity Conservation of Natural Ecosystems

Although biodiversity changes are important for the functioning of ecosystems and the welfare of people, as pointed out by Sala et al. (2000) there are several uncertainties regarding the protection of biodiversity per unit of land-use change, changes in atmospheric composition and future climate and in vegetation as well as regarding policies. The authors warn that hotspots of diversity (such as riparian corridors and coastal land margins) often coincide with centers of development, leading to high loss of diversity. It is also known that biodiversity in all biomes is sensitive to global changes and predictions of biodiversity change will require an integrated effort by climatologists, ecologists, social scientists, and policy makers to analyze the effects of future changes in the planet.

Nowadays it is more evident that restoration efforts may include the management of soil microbial communities including AMF networks (Hawkes et al. 2006, Grman and Suding 2009).Furthermore, AMF are sensitive to the order of host plant establishment, being important for plant species-dependent controls of AMF that may lead to alternative communities in the same environment. This is also related to biological invasions and climate change; that can alter plant community composition and phenology (Fitter and Fitter 2002, Suttle et al. 2007), confirming that attention must be paid to the stability of plant interactions with the AMF community (Hausmann et al. 2010).

Moreover, research must focus on the functional diversity of AMF (specialist or generalist) and how they affect plant fitness, as well as on the temporal scale of AMF turnover to establish priority actions due to their biological relevance (Hausmann et al. 2010).

In the studies by Pagano on the AM communities in the Brazilian dry forest and in riparian forest restoration experiments, the choice of tree species for restoration purposes seems to have great implications on the conservation of AMF species;thus, highly dependent tree hosts need to be selected over mycorrhizal- independent hosts. Moreover, in the studies coordinated by Cabello (Velazquez et al. 2008, 2010) in El Palmar National Park, Entre Ríosprovince, Argentina, ~78% of native plant species from scrublands, grassland, marsh and palm savannas dominated by *Butia yatay* presented AM colonization, and new AM species (spores) from the New World, such as *Acaulospora entreriana*, were found among 46 identified AMF.

Natural Vegetational Types in the Semiarid of Brazil and Argentina

In a comprehensive report, Morrone (2001 and references therein) described the biogeography of Latin America and the Caribbean, covering not only the flora and fauna but also the conservation level of each area when such information was available. Thus, as regards arid and semiarid environmental conditions, South America can be biographically depicted by the Neotropical Region, Subregion "Chaqueña" including Chaco and Monte Provinces within the Argentinean territory; and the Provinces of Caatinga, Cerrado and Chaco in Brazil. The "Andina" Region, Subregion "Páramo Puneña", includes Puna and Prepuna Provinces; the "Patagónica" Subregion, Patagonia Central and Patagonia "Subandina" Provinces along Argentina. The conservation status of all these South American arid and semiarid biogeographic Provinces is threatened by the expansion of agricultural frontiers, livestock, fire and wood collection (see Dinerstein et al. 1995 in Neiff 1996).

The weather of Subregion Chaqueña (Morrone 2001) or Domain Chaqueño (Cabrera and Willink 1980), Puna and Patagonia Central Provinces, is predominantly continental with low and moderate rainfall in summer. The vegetation of these harsh South American biogeographical Provinces is polymorphic and is represented by dry forests, deciduous forest, scrub, steppe, grasslands, etc.; in general, xerophytic vegetation predominates all over them.

Furthermore, each Province is characterized by its own altitudinal, climatic and floristic features (Cabrera and Willink 1980).

The semiarid region of Brazil (Caatinga formation) comprises a strip in north of Minas Gerais State, following the São Francisco River until south-easternState of Piauí. The characterization of the life-form flora and floristic for the State of Ceará was presented by Costa et al. (2007) and Araújo et al. (2011). The Caatinga formationpresented different physiognomic types, locally denominated as shrubby, woody, shrubby/ woody and park. In all physiognomic types of Caatinga, the richness of herbaceous stratum is high; however, there are only a few studies concerning the floristic, structure and population dynamics of the herbs (reviewed by Araújo et al. 2007). Most plants from the Caatinga enter a dormant phase during the major part of the year (dry season), while leaves, flowers, and the herbaceous vegetation grow in the short and sporadic rainy season (Rizzini 1997,Costa et al. 2007). Seasonal rainfall is in general considered the main factor influencing flowering and fruiting phenologies in dry forests; for a review see Pagano and Araújo (2011).

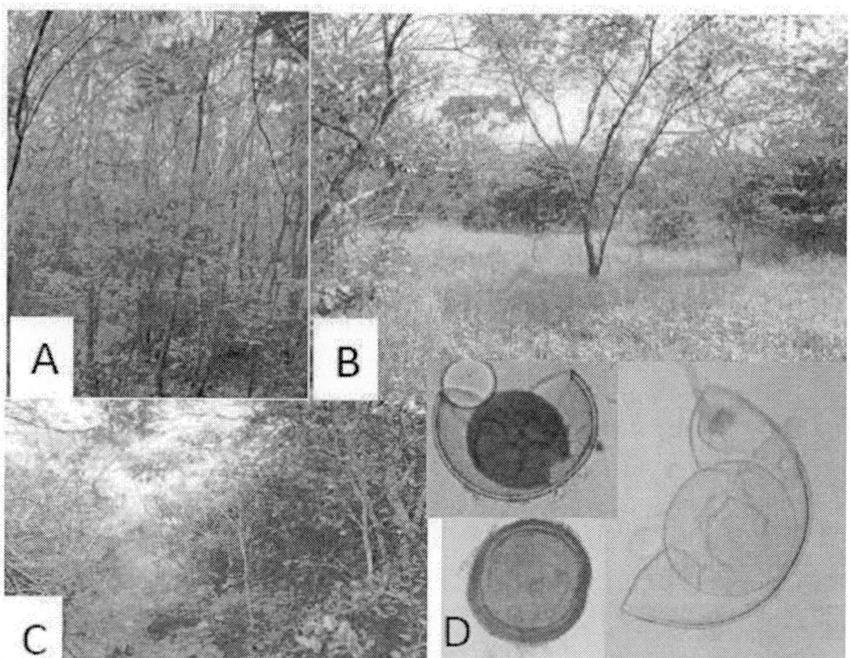

Figure 2. Physiognomic types of Brazil: (A) Dry forest; (B) Caatinga; (C) Carrasco and (D) AMF spores recovered.Photos by M. Pagano.

Woody Caatinga species attain a height of 30m and have intense leaf loss during the dry season (Prado 2003). Other vegetation types (Carrasco and Dry forest) from planalto of Ibiapaba, Ceará (Figure 2) benefit from the topoclimate and soil moisture conditions, which makes them functionally different. The soils in the Carrasco are arenosolsdcep and provide a lower water stress in this ecosystem due to a decrease in temperature and an increase in water availability in the soil (Araújo et al. 2011).

In the semiarid of Brazil the richness of herbaceous flora is approx. three times that of the woody stratum(Costa et al. 2007), and out of 1.102 species of trees and shrubs, 318 of them are endemic (reviewed by Pagano and Araújo 2011). The most common families are Euphorbiaceae, Leguminosae, Cactaceae and Anacardiaceae, and the species *Poincianella pyramidalis*(*Caesalpinia pyramidalis*), *Anadenanthera colubrina*, *Schinopsis brasiliensis* and *Myracrodruon urudeuva* present wide distribution (reviewed by Araújo et al. 2007), some of them being alsoincluded in the threatened or near threatened category of the official Brazilian endangered species list.

Space occupation of the Caatinga dominium by man and the exploration of its natural resources resulted in fragmented areas of several degrees of disturbance, only 16% of the total area being protected conservation unities. In intensively degraded areas, when plant cover is almost totally removed, the soils are degraded, the farming productive capacity has decreased and the social conditions of local populations are now deteriorated (reviewed by Araújo et al. 2007).

In arid, semiarid and tropical regions, almost a third of the world's irrigated land and half of total agricultural land cannot be used due to its high salinity. Each year, about 10 million hectares intended for agriculture lose their productivity. In this regard, arid and semi-arid areas account for 75% of the Argentinean surface (Pardo et al. 2002) and their salt content tends to increase due to agricultural practices.

Eighty percent of the Argentinean territory is used for agricultural activities, livestock and forestry; of the total country area (ca. 224.000.000 ha), 60.000.000 have suffered erosion and 650.000 ha are annually added to the total eroded areas, with varying degrees of erosion. These erosion processes are exacerbated in arid and semiarid zones, which constitute 75% of the country and contribute to 50% of agricultural production. A fact that cannot be ignored is that the anthropogenic impact on these environments exercised by traditional farming and livestock practices, inadequate management of natural and water resources and forest clearing, has contributed to the deterioration of native environments pushing them toward desertification. Taken together, these factors imply lower productivity and lead to the detriment of the living conditions and ultimately the poverty conditions of people and the abandonment of these areas (as reported by INTA, EEA and INTA AER Secantlás, Salta, Argentina).

Each one of the Argentinean arid and semiarid regions of Chaco, Monte, Puna and Patagonia areas have a distinctive species composition, although their vegetation encompasses in general xerophytic plants. A description of those vegetation types is as follows.

The Chaco vegetation is made up by xerophylic species forming deciduous dry forest, open scrublands, alternating with hard grass steppes; also, halophytic shrubs are widespread along this biogeographical province. The Angiosperm taxa characteristic of Chaco are the palms *Copernicia* and *Trithrinax*; Bromeliaceae, *Bromelia* and *Dickya*; Asteraceae, *Holocheilus* and *Panphalea*; Cactaceae of the genera *Opuntia*, *Cereus* and *Trichocereus*; the Zygophyllaceae, *Bulnesia*, *Larrea* and *Plectrocarpa*, Ulmaceae of the genus *Celtis*; Anacardiaceae genera *Aspidosperma* (quebrachos), *Lithraea,Schinopsis* and *Schinus*, and the Fabaceae as *Acacia* and *Geophrea* among others. At present, this phytogeographic province is highly degraded by intensive forestal exploitation and livestock, and great areas of the primitive forests and even the herb stratum have disappeared, giving place to thorn bushes, terrestrial Bromeliaceae and Cactaceae. Thus, in the degraded land invasion by *Larrea divaricata* (Zygophyllaceae) is commonly found, which is a typical species of the Monte (Cabrera 1976, Cabrera and Willink 1980, Morrone 2001).

Throughout the Monte phytogeographic province plant communities such as xerophilic, sammophilic and halophilic shrub steppe prevail; furthermore, scant forest of mimosoid leguminose and *Salix* are found along river sides or in humid areas. The vegetation of the Monte is dominated by Zygophyllaceae bushes of *Larrea* (mixed with shrubs of *Prosopis* genera), *Bulnesia*, *Plectocarpa*, and *Zuccagnia*. Cactaceae are abundant, *Denmoza*, *Pterocactus*, *Setiechinopsis* and *Tephrocactus* being representative taxa; besides, genera such as *Prosopis*, *Atamisquea*, *Cercidium*, *Bulnesia*, *Bougainvillea*, and *Cordalia* among others, are also present in the Chaco region. The "Jarillal" is the most representative community of Monte, which mainly includes *Larrea divaricata*, *L. cuneifolia*, and *L. nitida* developed in shallow vales and plains with sandy or rocky-sandy soils. Moreover, the vegetation is constituted by annual, geophytes, succulent (e. g. Cactaceae, Chenopodiaceae and Portulaceae) and perennial species, resistant to drought. The members of the genera *Monttea*, *Bulnesia*, *Cassia* and *Baccharis* are the perennial plant commonly growing in Monte and the dominant species are *Allenrolfea vaginata*, *Atriplex lampa*, *Baccharis salicifolia*, *Bougainvilleaspeciosa*, *Bulnesia retama*, *B. schikendantzii*, *Cassia rigida*, *Cercidium praecox*, *Fabiana patagonica*, *Larrea divaricata*, *L. cuneifolia*, *L. nitida*, *Mimosa ephedroides*, *Montea aphylla*, *Plectocarpa rougessi*, *Portulaca* spp., *Prosopis alpataco*, *P. chilensis*, *P. flexuosa*, *P. torquata*, *Proustia cuneifolia*, *Tessaria dodoneaefolia* and *Zuccagnia punctata* (Cabrera 1971, 1976, Cabrera and Willink 1980, Morrone 2001).

The flora from Patagonia is constituted by steppes of shrubs, chamaephytes and herbaceous species.Sammophylic and halophylic steppes with predominance of cushion plants are also found. In the western wet region of Patagonia, the graminoid steppes are frequent. *Ameghinoa*, *Pantacantha*, *Benthamiella*, *Lepidophyllum* are endemic genera. Asteraceae is the most important family represented by some species of *Senesio*, *Nassauvia*, *Perezia*, *Chuquiraga*, followed by Poaceae, Fabaceae, Verbenaceae and Solanaceae. The angisoperm taxa that characterize this subregion are *Azorella ameghinoi*, *A. montana*, *A. patagonica* and *Oligocladus* (Apiacea); *Ameghinoa*, *Burkartia*, *Duseniella*, *Leucheria achillaeifolia*, *L. lithospermifolia*, *Naussavia ameghinoi*, *N. fuegiana* and *N. juniperina* (Asteraceae); *Delphinophitum*, *Pterygiosperma*, *Sarcodraba*, *Skottsbergianthus* and *Trichotolium* (Brassicaceae); *Maihuenia* (Cactaceae) and *Aonikena* (Euphorbiaceae). In Patagonia Central, *Agrostis pyrogea*, *Anarthrophyllum rigidum*, *Chuquiraga aurea*, *C. avellanedae*, *Colliguaya integerrima*, *Distichlis scoparia,D. spicata*, *Festuca argentina*, *F. pallescens*, *Haplopappus pectinatus*, *Juncus leuserii*, *Junellia tridens*, *Mulinum spinosum*, *Nassauvia axilaris*, *N. glomerulosa*, *Poaligularis*, *Schinus marchandii*, *Senecio bracteolatus*, *S. filaginoides*, *Stipa* spp. and *Trevoa patagonica* integrate the dominant species (Cabrera 1971, 1976, Cabrera and Willink 1980). The taxa characteristic of Central Patagonia are *Chuquiraga aurea, Leucheria floribunda, L. gilliesii, L. rosea, L. runcinata, L. scrobiculata, Nassauvia uniflora and N. ruizii* (Morrone 2001).

The floristic of Puna is characterized by angiosperm taxa such as *Chuquiraga atacamensis* and *C. kuscheli*, Asteraceae; *Epibolium fragile*, *E. pedicellare*, *Fuchsia austromontana*, *F. bolivianus* and *F. simplicaulis*, Onagraceae. Analyses of panbiogeographic, biogeographic, and pasimony of Puna's endemism of vascular plants and insects has shown the close Puna-Andean Paramo relationship (Morrone 2001 and references therein). The dominant vegetation of Puna consists of steppes or shrublands of low bushes, 40-150 cm heigh, among which the ground is bare. During the summer, there are many annual and perennial species with underground organs that re-sprout during the rainy season. Furthermore, there are also grasslands in wetter soils and scant "queñoa" (*Polylepis tomentella*) forests in mountainsides and ravines. It is important to note that *P. tomentella* is the sole tree species in the region. Throughout the "wet Puna" (annual precipitation higher than 400 mm), predominant species are *Parastrephia lepidiphylla* and P. *phylicaephola* ("tola"); on the other hand, the "dry Puna" species are *Baccharis boliviensis, Fabiana densa, Adesmia horridiuscula, Acantholippia hastulata, Nardophyllum armatum* as common species of bushes. Some of the endemic genera are *Parastrephia*, *Lampaya*, *Chersodoma*, *Eustiphiopsis*, *Anthobryum*, *Oreocereus*, *Chiliotrichiopsis*, *Lophopappus* and *Urmenetea.*

Figure 3. Some phytogeographic types of semiarid and arid vegetation in Argentina. (A) Monte; (B) Chaco; (C) Jarillal, Monte; (D) Puna.Photos by M. A. Lugo.

In wet soils nearby water tables, rivers or lakes, *Cortaderia speciosa*, *Satureja parvifolia* and *Baccharis grisebachii* are found; in salty soils also *Festuca scirpifolia* and, among sandy-wet soil *Pennisetum chilense* are found (Cabrera and Willink 1980). Martínez Carretero (1995) analyzed 107 taxa of Puna that represented 16.5% from the 648 species reported by Cabrera in 1957 (Martínez Carretero 1995 and references therein). These Puna's taxa were grouped according to localities, species and phytosociological features. Thus, four grouping clades discriminated showed the Jujeño district to be characterized by the presence of *Mutisia saltensis*, *Chersodoma argentina*, *Stipa arcuata*, *Opuntia* cfr. *tilcarensis*, *Nototriche sarmentosa*, etc; furthermore, the Jujeño district coincides with part of the sampling area of the data discussed in this chapter. There are few studies recording the phytosociology of the Puna; however, for Jujuy, Ruthsatz (1977) reported 119 species and, among them, numerous associations and sub-associations were established, some related to soil features, and others to altitudinal variation. Thus, Ruthsatz (1977) mentioned associations of *Junellia asparagoides-Proustia cuneifolia* on slopes, *Fabiana densa-Baccharis boliviensis* in the plains, *Hypseocharis-Portulacca perennis* in salt marshes, *Parasthrephia lepidophylla* in deep sandy soils of riverbed and of *Fabiana denudata-Baccharis boliviensis* var. *latifolia* in stony soils, among others. Moreover, the Jujeño district is in agreement with the dry Puna ("Puna seca") as described by Troll in 1959 (Martínez Carretero 1995 and references therein).

Native plants and their mycorrhizal symbiosis have been analyzed by Argentine researchers throughout arid and semiarid phytogeographic areas (Figure 3). Studies of bacterial communities associated to native species are scant in those areas. One of the studies available is our analysis of the rhizospheric bacterial communities and mycorrhizas associated to native grasses of Puna, which was carried out using a combination of cultivation-dependent and independent fingerprinting techniques. The results will be discussed later in this chapter.

Mycorrhization in Endangered Plants

As nearly 3% of the described plant species of the planet are either threatened or are classified into different categories of threat (critical, endangered, and vulnerable) and this value is rising continually (IUCN 2006), the potential role of symbiont microorganismsin protecting endangered plants is nowadays in focus. Recent reports such as by Bothe et al. (2010) in Poland emphasizethe potential role of AMFin protecting endangered plants, pointing that AMF populations related to those species have not been investigated, which probably supports the thesis of AMF being unique to those sites. Additionally, Fuchs and Haselwandter (2004) began to study the mycorrhization of endangered plants (Table 1), considering the dark septate endophytes (DSE).

DSE (explained below) are found in several plants, mostly shrubs but also trees, herbaceous and gramineous species and usually in agronomic plants (e. g. Barrow and Aaltonen 2001, Barrow 2003, Chaudhry et al. 2005, Likar et al. 2008, Newsham et al. 2009, Newsham 2011, Stevens et al. 2010). With regard to DSE, Fuchs and Haselwandter (2004) found two plant species also regularly colonized by AMF and DSE; other plant species, however, were rarely colonized by AMF and DSE or not colonized by DSE at all in wetland sites in Salzburg.

In recent years, it is possible to observe a worldwide research trend toward including DSE in numerous research projects, particularly in extreme environments, in order to examine and use them for conservation and restoration projects (Figure 4).

In the semiarid of Brazil, two tree species (*Mimosa caesalpiniifolia* and *Cordia oncocalyx*) considered as threatened, presented AM symbiosis (Pagano and Araújo 2010), both in the dry and wet seasons, though higher root colonization was seen in the dry period (Pagano, unpublished data).

Table 1. Database survey conducted in April 2011 (SCOPUS) for journal articles dealing with dark septate endophytes (DSE) and endangered plant species

Key words	Number of journal articles
Dark septate endophytes	160
Root dark septate endophytes	138
Root dark septate endophytes + AMF	23
Root dark septate endophytes + Endangered plant species	9
Root dark septate endophytes + South America	14
Root dark septate endophytes + Endangered plant species + South America	1

AMF = arbuscular mycorrhizal fungi.

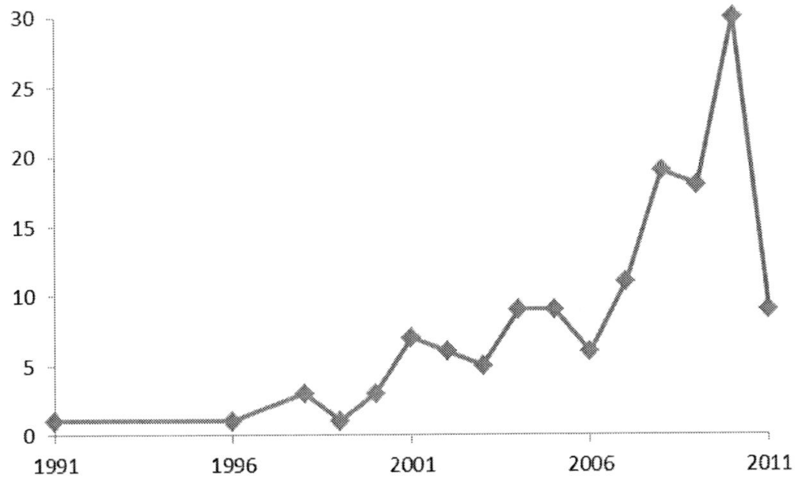

Figure 4. Number of papers on DSE published annually since 1991, included in the SCOPUS. Only those papers that had appeared in the SCOPUS by April 2011 are included (i.e. > 33% of the total for that year).

Table 2. Summary of evidence on arbuscular mycorrhizal fungi in some threatened plants from South America

Source	Location/ forest type	Plant species	Mycorrhizal or DSE colonization[#]
Pagano et al. (2010)	Native preserved caatinga Serra das Almas, Ceará State, Brazil	*Cordia oncocalix, Mimosa caesalpiniifolia*	60 and 29%
Fracchia et al. (2009)	Chaco Serrano Woodland community on the west hillside of the Sierras Chicas, Punilla Department, Argentina	*Buddleja cordobensis, Aa achalensis*	Medium, DSE
Pagano et al. (2008, 2010)	Dry forest, Minas Gerais State, Brazil	*Schinopsis brasiliensis*	50%
Amorim et al. (2004)	Feira de Santana, Bahia State, Brazil	*Myracroduon urundeuva*	AM

[#]Maximal AM colonization reported for each plant species, respectively, Medium= AM colonization class medium; DSE = dark septate endophytes.

Table 3. Mycorrhizal status and distribution of some native species in the semiarid of Brazil

Family	Plant species	Reference	ES†	Distribution
Mimosaceae	*Anadenanthera peregrina*	AM[1]	-	Woody caatinga
	Enterolobium contortisiliquum	AM[2]	-	Dry forest
	Mimosa caesalpiniifolia†	AM[4]	-	Caatinga
	Mimosa tenuiflora	AM[4]	-	Caatinga
	Piptadenia moniliformis	NM[4]	-	Dry forest
Fabaceae	*Anadenanthera colubrina*	AM[6]	-	Caatinga
Caesalpiniaceae	*Pointianella bracteosa*	AM[4]	-	Caatinga
	Bauhinia cf. *pulchella*	AM[4]	-	Dry forest
	Bauhinia acuruana	AM[4]	-	Carrasco
	Copaifera langsdorfii	AM[4]	-	Dry forest
Euphorbiaceae	*Croton blanchetianus*	AM[4]	-	Caatinga
	Gymnanthes sp.	AM[4]	-	Dry forest
	Croton argirophyloides	NM[4]	-	Dry forest
	Sapium lanceolatum	AM[4]	-	Carrasco
Apocynaceae	*Aspidosperma pyrifolium*	AM[4]	-	Caatinga
	Aspidosperma subincanum	AM[4]	-	Dry forest
Boraginaceae	*Cordia oncocalyx*†	AM[4]	+	Caatinga
Bignoniaceae	*Handroanthus heptaphyllus*	AM[2]	-	Dry forest
Malpighiaceae	*Byrsonima garderiana*	AM[4]	-	Carrasco
Myrtaceae	*Eugenia* aff. *dysenterica*	AM[4]	-	Carrasco
	Eugenia sp.	AM[4]	-	Carrasco
Flacourtiaceae	*Xylosma ciliatifolium*	AM[4]	-	Dry forest
Leguminosae	*Plathymenia reticulata*	AM[2,3]	-	Dry forest
Rubiaceae	*Zanthoxylum stelligerum*	NM[4]	-	Carrasco
Arecaceae	*Copernicia prunifera*	ND	-	Flood plains, Northeast region
Anacardiaceae	*Schinopsis brasiliensis*†	AM[2]	+	Dry forest
	Anacardium occidentale	AM [1,5]	-	Domesticated, Northeast region

Note: Indicated are the families. AM = Arbuscular mycorrhizal colonization report. [1]Wang and Qiu (2006); [2]Pagano et al. (2010); [3]Pagano et al. (2009); [4]This study; [5]Weber et al. (2004); [6]Patreze and Cordeiro (2004) and Zangaro et al. (2003); DSE= dark septate endophytes colonization; ND = Not determined; NM= Not mycorrhizal. †Endangered or near threatened plant species.

Among the woody species cataloged mostly for use in construction or for fuel, *Myracrodruon urundeuva*, *Schinopsis brasiliensis* , and *Anadenanthera colubrina* were among the prominent ones due to their local importance and multiplicity of uses, in spite of the fact that the first two trees are included in Brazilian lists of threatened species (Lucena et al. 2007). Mycorrhizal associations established between AMF, DSE and some native threatened plants from South America are summarized in Table 2.

Mycorrhizal associations established between AMF, DSE and native threatened plants from Brazil are summarized in Table 3.

The studies of mycorrhizal association among Argentinean arid and semiarid regions showed AM as the most widespread mycorrhizal type, though only one ericoid mycorrhiza had been represented (Fontenla et al. 2001, Fracchia et al. 2009, Lugo et al. 2005, 2008, 2009, Urcelay 2002). Arbuscular mycorrhizal root colonization values were in general low as well as AMF species diversity (Table 4, 5); however, maximal values of root colonization were recorded in few plant species. The conservation status of host plants differed between species, such as *Astrocedrus chilensis* and *Schinopsis hankeana* are included in the Red List (Table 5). Arbuscular mycorrhizal colonization, AMF spore diversity, and DSEs were studied in the

Table 4. Summary of evidence on arbuscular mycorrhizal fungi in the arid and semi-arid region of Argentina, South America

Source	Location/ forest type	Spore number[1]	AMF Species Richness[2]	Mycorrhizal colonization[3]
This study	Puna, North West of Argentina	ND	ND	68%
Lugo et al. (2009)	Sierras de las Quijadas National Park, San Luis	ND	ND	100%
Fracchia et al 2009	Chaco Serrano Woodland, central Argentina	ND	ND	>30%
Lugo et al. (2008)	Puna, North West of Argentina	107	10	ND
Lugo et al. (2005)	Jarillal, Chaco Phytogeographical Province, Córdoba	169	7	23%

[1]Maximal Spore number 100g^{-1} soil, [2]Maximal AMF Species Richness, [3]Maximal AM colonization reported, ND = not determined in the study.

roots and rhizospheric soil of Poaceae native from Puna (see below). The symbiotic status of the Argentinean native plants summarized in Tables 4 and 5 is illustrated in Figure 5.

In Argentina, both vulnerable species *Astrocedrus chilensis* and *Schinopsis hankeana* inhabit areas of Patagonia with higher precipitation than the arid and semiarid region focused on this chapter. However, it is important to notice that *Astrocedrus chilensis* is reported as AM (Fontenla et al. 1991); besides, their forest mycorrhizal community dynamics has already been studied further considering its pathogenic relationships (Fontenla et al. 1998, 2001). At present, *A. chilensis* and *S. hankeana* are included in programs of the "Global Restoration Network", an international conservation project of the Society for Ecological Restoration International (http://www.globalrestorationnetwork.org/grn/estudios-de-caso/).

Table 5. Mycorrhizal status of some native species in Argentina, South America

Family	Plant species	Reference	ES[†]	Distribution
Acanthaceae	*Dicliptera squarrosa*	AM-	-	Chaco
	Stenandrium dulce	AM-	-	Chaco
Alliaceae	*Nothoscordum gracile*	AM-	-	Chaco
Amaranthaceae	*Amaranthus viridis*	NM[1]	-	Chaco
	Gomphrena pulchella	AM-	-	Chaco
Amaryllidaceae	*Rhodophiala mendocina*	NM-ND[2]	-	Patagonia
Anacardiaceae	*Lithraea molleoides*	AM-	-	Chaco
	Schinopsis haenkeana	AM-	†	Chaco
	Schinus areira	AM-	-	Chaco
	Schinus fasciculata	AM-	-	Chaco
	Schinus patagonicus	AM-ND[2]	-	Patagonia
Apiaceae	*Azorella ameghinoi*	AM-ND[2]	-	Patagonia
	Eryngium horridum	AM-	-	Chaco
	Eryngium paniculatum	AM-	-	Chaco
	Mulinum microphyllum	AM-ND[2]	†	Patagonia
	Mulinum spinosum	AM-ND[2]	-	Patagonia
	Osmorhiza chilensis	AM-ND[2]	-	Patagonia
Apocynaceae	*Amblyopetalum coccineum*	AM[1]	-	Chaco
	Aspidosperma quebracho-blanco	AM-	-	Chaco
Arecaceae	*Trithrinax campestris*	AM-	-	Chaco
Asclepiadaceae	*Asclepias mellodora*	AM-	-	Chaco

Family	*Plant species*	Reference	ES†	Distribution
Asteraceae	*Achyrocline satureoides*	AM-	-	Chaco
	Achyrocline tomentosa	AM-	-	Chaco
	Agoseris coronopifolia	AM-ND[2]	-	Patagonia
	Baccharis articulata	AM[1]	-	Chaco
	Baccharis obovata	AM-ND[2]	-	Patagonia
	Bidens pilosa	AM-	-	Chaco
	Chaptalia sinuata	AM-	-	Chaco
	Chuquiraga oppositifolia	AM-ND[2]	-	Patagonia
	Eupatorium arnottianum	AM[1]	-	Chaco
	Flourensia campestris	AM-	†	Chaco
	Flourensia oolepis	AM-	†	Chaco
	Gaillardia megapotamica	AM-	-	Chaco
	Galinsoga parviflora	AM-	-	Chaco
	Grindelia cabrerae	AM-	†	Chaco
	Haploppapus glutinosus	AM-ND[2]	-	Patagonia
	Heterothalamus alienus	AM-	-	Chaco
	Hypochaeris aff. incana	AM-ND[2]	-	Patagonia
	Mutisia decurrens	AM-ND[2]	-	Patagonia
	Mutisia spinosa	AM-ND[2]	-	Patagonia
	Nassauvia aculeata	AM-ND[2]	-	Patagonia
	Nassauvia glomerulosa	AM-ND2	-	Patagonia
	Parthenium hysterophorus	AM-	-	Chaco
	Perezia recurvata	AM-ND2	-	Patagonia
	Senecio bracteolatus	AM-ND2	-	Patagonia
	Senecio gymnocaulos	AM-ND2	†	Patagonia
	Senecio neaei	AM-ND2	-	Patagonia
	Senecio sericeo-nitens	AM-ND2	-	Patagonia
	Senecio subumbellatus	AM-ND2	-	Patagonia
	Stevia satureiifolia	AM-	-	Chaco
	Tagetes minuta	AM-	-	Chaco
	Tragopogon dubius	AM-ND2	-	Patagonia
	Trichocline reptans	AM-	-	Chaco
	Trichocline sinuata	AM-	-	Chaco
	Triptilion achilleae	NM-ND2	-	Patagonia
	Vernonia mollissima	AM-	-	Chaco
	Vernonia nudiflora	AM-	-	Chaco
	Xanthium cavanillesii	AM-	-	Chaco
	Zinnia peruviana (L.)	AM-	-	Chaco
Berberidaceae	*Berberis aff. buxifolia*	AM-ND2	-	Patagonia
	Berberis heterophylla	NM-ND2	-	Patagonia
	Berberis ruscifolia	AM1	-	Chaco

Table 5. (Continued)

Family	Plant species	Reference	ES†	Distribution
Boraginaceae	*Heliotropium amplexicaule*	NM1	-	Chaco
	Plagiobothrys verrucosus	AM-ND2	-	Patagonia
Brassicaceae	*Erophila verna*	NM-ND2	-	Patagonia
	Rorippa nasturtium-aquaticum	NM-ND2	-	Patagonia
	Thlaspi magellanicum	NM-ND2	-	Patagonia
Bromeliaceae	*Bromelia urbaniana*	NM[3]	-	Monte,
	Deuterocohnia longipetala	AM-DSE[3]	-	Monte,
	Dyckia floribunda	AM-DSE[3]	-	Monte,
	Dyckia velascana	AM-DSE[3]	†	Monte,
	Puya spathacea	AM[1]	-	Chaco
	Tillandsia angulosa	AM-DSE[3]	-	Monte,
	Tillandsia bryoides	DSE[3]	-	Monte,
	Tillandsia castellanii	DSE[3] †	-	Monte,
	Tillandsia rectangula	DSE[3]	-	Monte,
	Tillandsia xiphioides	DSE[3]	-	Monte,
Buddlejaceae	*Buddleja cordobensis*	AM-DSE[1]	†	Chaco
Cactaceae	*Acanthocalycium spiniflorum*	AM-DSE[1]	-	Chaco
	Gymnocalycium monvillei	AM-DSE[1]	†	Chaco
	Maihuenia patagonica	NM-ND[2]	-	Patagonia
	Opuntia sulphurea	AM-DSE[1]	-	Chaco
	Trichocereus candicans	AM-DSE[1]	†	Chaco
Calyceraceae	*Boopis australis*	NM-ND[2]	-	Patagonia
	Nastanthus aff. patagonicus	NM-ND[2]	†	Patagonia
Cannaceae	*Canna indica*	AM-DSE[1]	-	Chaco
Caryophyllaceae	*Arenaria serpyllifolia*	NM-ND[2]	-	Patagonia
	Cerastium arvense	NM-ND[2]	-	Patagonia
	Holosteum umbellatum	NM-ND[2]	-	Patagonia
	Scleranthus annuus	NM-ND[2]	-	Patagonia
Celastraceae	*Maytenus chubutensis*	AM-ND[2]	-	Patagonia
Celtidaceae	*Celtis pallida*	AM-DSE[1]	-	Chaco
	Celtis tala	AM-DSE[1]	-	Chaco
Chenopodiaceae	*Chenopodium ambrosioides*	NM[1]	-	Chaco
Cleomaceae	*Cleome aculeata*	NM[1]	-	Chaco
Commelinaceae	*Commelina erecta*	NM[1]	-	Chaco
Convolvulaceae	*Dichondra microcalyx*	AM-DSE[1]	-	Chaco
	Dichondra sericea	AM-DSE[1]	-	Chaco
	Ipomoea purpurea	AM[1]	-	Chaco
Cruciferaceae	*Lepidium aletes*	NM[1]	-	Chaco

Family	*Plant species*	Reference	ES†	Distribution
Cupressaceae	*Austrocedrus chilensis*	AM-ND[2]	†	Patagonia
Cyperaceae	*Cyperus entrerianus*	AM[1]	-	Chaco
	Carex fuscula	AM-ND[2]	-	Patagonia
	Carex gayana	NM-ND[2]	-	Patagonia
	Carex macloviana	NM-ND[2]	-	Patagonia
	Eleocharis aff. pachycarpa	NM-ND[2]	-	Patagonia
Ephedraceae	*Ephedra frustillata*	NM-ND[2]	-	Patagonia
	Ephedra triandra	AM-DSE[1]	-	Chaco
Equisetaceae	*Equisetum bogotense*	AM-ND[2]	-	Patagonia
	Equisetum giganteum	AM[1]	-	Chaco
Euphorbiaceae	*Acalypha communis*	AM[1]	-	Chaco
	Croton sarcopetalus	AM-DSE[1]	-	Chaco
	Euphorbia portulacoides	AM[1]	-	Chaco
	Euphorbia serpens	AM[1]	-	Chaco
	Stillingia patagonica	AM-ND[2]	†	Patagonia
Fabaceae	*Acacia atramentaria*	AM-DSE[1]	-	Chaco
	Acacia caven	AM-DSE[1]	-	Chaco
	Acacia praecox	AM-DSE[1]	-	Chaco
	Acacia visco	AM-DSE[1]	-	Chaco
	Adesmia aff. aucaensis	AM-ND[2]	-	Patagonia
	Adesmia boronioides	AM-ND[2]	-	Patagonia
	Adesmia aff. corymbosa	AM-ND[2]	-	Patagonia
	Adesmia volckmanni	AM-ND[2]	-	Patagonia
	Anarthrophyllum rigidum	AM-ND[2]	-	Patagonia
	Caesalpinia gilliessi	AM-DSE[1]	-	Chaco
	Collaea argentina	AM-DSE[1]	-	Chaco
	Geoffroea decorticans	AM-DSE1	-	Chaco
	Lathyrus magellanicus	AM-ND2	-	Patagonia
	Lotus pedunculatus	AM-ND2	-	Patagonia
	Prosopis alba	AM-DSE1	-	Chaco
	Prosopis nigra	AM-DSE1	-	Chaco
	Senna aphylla	AM-DSE1	-	Chaco
	Sophora linearifolia	AM-DSE1	†	Chaco
	Vicia nigricans	AM-ND2	-	Patagonia
Geraniaceae	*Erodium cicutarium*	AM-ND2	-	Patagonia
	Geranium magellanicum	AM-ND2	-	Patagonia
Gunneraceae	*Gunnera magellanica*	AM-ND2	-	Patagonia
Hydrophyllaceae	*Phacelia secunda*	AM-ND2	-	Patagonia
Iridaceae	*Olsynium junceum*	AM-ND2	-	Patagonia
	Sisyrinchium chilense	AM-DSE1	-	Chaco
Lamiaceae	*Hedeoma multiflora*	AM-DSE1	-	Chaco

Table 5. (Continued)

Family	*Plant species*	Reference	ES†	Distribution
Ledocarpaceae	*Balbisia gracilis*	AM-ND2	-	Patagonia
Loasaceae	*Caiophora aff. patagonica*	AM-ND2	-	Patagonia
Lythraceae	*Cuphea glutinosa*	AM-DSE1	-	Chaco
	Heimia salicifolia	AM-DSE1	-	Chaco
Malphigiaceae	*Janusia guaranitica*	AM-DSE1	-	Chaco
Malvaceae	*Abutilon grandifolium*	AM1	-	Chaco
	Gaya parviflora	AM-DSE1	-	Chaco
	Mediolastrum gilliesii	AM-DSE1	-	Chaco
	Sphaeralcea cordobensis	AM-DSE1	†	Chaco
	Tarasa humilis	AM-ND2	-	Patagonia
Onagraceae	*Clarkia tenella*	AM-ND[2]	-	*Patagonia*
	Oenothera odorata	AM-ND[2]	-	*Patagonia*
	Oenothera affinis	AM-DSE[1]	-	*Chaco*
Oxalidaceae	*Oxalis conorrhiza*	AM-DSE[1]	-	*Chaco*
Papaveraceae	*Argemone subfusiformis*	NM[1]	-	*Chaco*
Passifloraceae	*Passiflora caerulea*	AM-DSE[1]	-	*Chaco*
Plantaginaceae	*Plantago australis*	AM-DSE[1]	-	*Chaco*
	Plantago lanceolata	AM-ND[2]	-	*Patagonia*
	Plantago tomentosa	AM-DSE[1]	-	*Chaco*
Plumbaginaceae	*Armeria maritima*	AM-ND[2]	-	*Patagonia*
Poaceae	*Aristida adscensionis*	AM-DSE[5]	-	*Puna*
	Aristida asplundii	AM-DSE[5]	-	*Puna*
	Bouteloua barbata	AM-DSE[5]	-	*Puna*
	Bromus catharticus	AM-DSE[5]	-	*Puna*
	Bromus aff. setifolius	AM-ND[2]	-	*Patagonia*
	Bouteloua simplex	AM-DSE[5]	-	*Puna*
	Cenchrus incertus	AM-DSE[1]	-	*Chaco*
	Cortaderia selloana	AM-DSE[1]	-	*Chaco*
	Danthonia annableae	AM-DSE[5]	-	*Puna*
	Distichlis scoparia	AM-ND[2]	-	*Patagonia*
	Eragrostis nigricans	AM-DSE[5]	-	*Puna*
	Festuca pallescens	AM-ND[2]	-	*Patagonia*
	Holcus lanatus	AM-ND[2]	-	*Patagonia*
	Hordeum comosum	NM-ND[2]	-	*Patagonia*
	Hordeum murinum	NM-ND[2]	-	*Patagonia*
	Jarava plumosula	AM-DSE[5]	-	*Puna*
	Neobouteloua lophostachya	AM-ND[4]	-	*Monte*
	Poa lanuginosa	AM-ND[2]	-	*Patagonia*
	Poa ligularis	AM-ND[2]	-	*Patagonia*

Family	*Plant species*	Reference	ES†	*Distribution*
	Polypogon interruptus	AM-DSE[5]	-	*Puna*
	Sporobolus pyramidatus	AM-ND[4]	-	*Monte*
	Stipa neaei	NM-ND[2]	-	*Patagonia*
	Stipa poeppigiana	AM-ND[2]	-	*Patagonia*
	Stipa speciosa	AM-ND[2]	-	*Patagonia*
	Rytidosperma virescens	AM-ND[2]	-	*Patagonia*
	Trichloris crinita	AM-ND[4]	-	*Monte*
	Vulpia australis	AM-ND[2]	-	*Patagonia*
	Vulpia bromoides	AM-ND[2]	-	*Patagonia*
Polemoniaceae	*Collomia linearis*	AM-ND[2]	-	*Patagonia*
Polygonaceae	*Ruprechtia apetala*	AM-DSE[1]	-	*Chaco*
Portulacaceae	*Montiopsis gayana*	NM[2]	-	*Patagonia*
	Portulaca grandiflora	NM[1]	-	*Chaco*
Pteridaceae	*Adiantum raddianum*	AM[1]	-	*Chaco*
	Argyrochosma nivea	AM[1]	-	*Chaco*
Ranunculaceae	*Anemone multifida*	AM-ND[2]	-	*Patagonia*
	Clematis montevidensis	AM-DSE[1]	-	*Chaco*
	Thalictrum decipiens	AM-DSE[1]	-	*Chaco*
Rhamnaceae	*Colletia spinosissima*	AM-DSE[1]	-	*Chaco*
	Discaria articulata	AM-ND[2]	-	*Patagonia*
	Discaria trinervis	AM-ND[2]	-	*Patagonia*
Rosaceae	*Acaena magellanica*	AM-ND[2]	-	*Patagonia*
	Acaena myriophylla	AM-DSE[1]	-	*Chaco*
	Acaena pinnatifida	AM-ND[2]	-	*Patagonia*
	Acaena aff. platyacantha	AM-ND[2]	-	*Patagonia*
	Fragaria chiloensis	AM-ND[2]	-	*Patagonia*
	Geum magellanicum	AM-ND[2]	-	*Patagonia*
	Kageneckia lanceolata	AM-DSE[1]	-	*Chaco*
	Tetraglochin caespitosum	AM-ND[2]	-	*Patagonia*
Rubiaceae	*Borreria densiflora*	AM[1]	-	*Chaco*
	Galium latoramosum	AM-ND[2]	-	*Patagonia*
	Galium richardianum	AM-ND2	-	*Patagonia*
	Oreopolus glaciali	AM-ND2	-	*Patagonia*
Rutaceae	*Fagara coco*	AM-ND2	-	*Patagonia*
Salicaceae	*Salix humboldtiana*	AM-DSE1	-	*Chaco*
Santalaceae	*Arjona tuberosa*	NM-ND2	-	*Patagonia*
	Jodinia rhombifolia	AM1	-	*Chaco*
	Myoschilos oblongum	NM-ND2	-	*Patagonia*
	Quinchamalium chilense	AM-ND2	-	*Patagonia*

Table 5. (Continued)

Family	*Plant species*	Reference	ES†	*Distribution*
Saxifragaceae	*Ribes cucullatum*	AM-ND2	-	*Patagonia*
Schizaeaceae	*Anemia tomentosa*	AM1	-	*Chaco*
Scrophulariaceae	*Calceolaria polyrrhiza*	AM-ND2	-	*Patagonia*
Solanaceae	*Cestrum parqui*	AM-DSE1	-	*Chaco*
	Fabiana imbricata	AM-ND2	-	*Patagonia*
	Nicotiana glauca	AM-DSE1	-	*Chaco*
	Petunia axillaria	AM1	-	*Chaco*
	Salpichroa origanifolia	AM1	-	*Chaco*
	Solanum palinacanthum	AM1	-	*Chaco*
	Solanum sisymbrifolium	AM-DSE1	-	*Chaco*
Turneraceae	*Turnera sidoides*	AM-DSE1	-	*Chaco*
Valerianaceae	*Valeriana carnosa*	AM-ND2	-	*Patagonia*
	Nicotiana longiflora	AM-DSE[1]	-	*Chaco*
	Nierembergia linariaefolia	AM-DSE[1]	-	*Chaco*
Verbenaceae	*Aloysia gratissima*	AM-DSE[1]	-	*Chaco*
	Glandularia dissecta	AM[1]	-	*Chaco*
	Glandularia peruviana	AM[1]	-	*Chaco*
	Junellia succulentifolia	AM-ND[2]	†	*Patagonia*
	Lippia turbinata	AM-DSE[1]	-	*Chaco*
	Verbena litoralis	AM-DSE[1]	-	*Chaco*
Violaceae	*Viola aff. cotyledon*	AM-ND[2]	-	*Patagonia*
	Viola maculata	AM-ND[2]	-	*Patagonia*
Zygophyllaceae	*Larrea divaricata*	AM-ND[4]	-	*Monte*

Note: Indicated are the families. AM = Arbuscular mycorrhizal colonization report. DSE= dark septate endophytes colonization; NM= Not mycorrhizal; ND = Not determined; [1]Fracchia et al. (2009); [2]Fontenla et al. (2001); [3]Lugo et al. (2009); [4]Lugo et al. (2005); [5]This study. [†]Endangered or near threatened plant species: includes species belonging to category 3, 4 or 5 according http://www.lista-planear.org and vulnerable species according http://www.ambiente.gov.ar.

DARK SEPTATE ENDOPHYTES IN ROOTS OF PLANTS

Since the first 1917 report of dark septate hypha from a fungal symbionts with negative effect on its host, living inside root tissues (endophytes) and often occurring together with other mycorrhizal associations (as AM, EM, ericoid and orchid mycorrhizas), increasing and more accurate information is available about them. However, dark septate endophytes (DSE) function as parasitic or mutualistic are still controversial, DSE potential relationship to hosts being considered as a continuum. Nutritional capabilities of DSE fungi

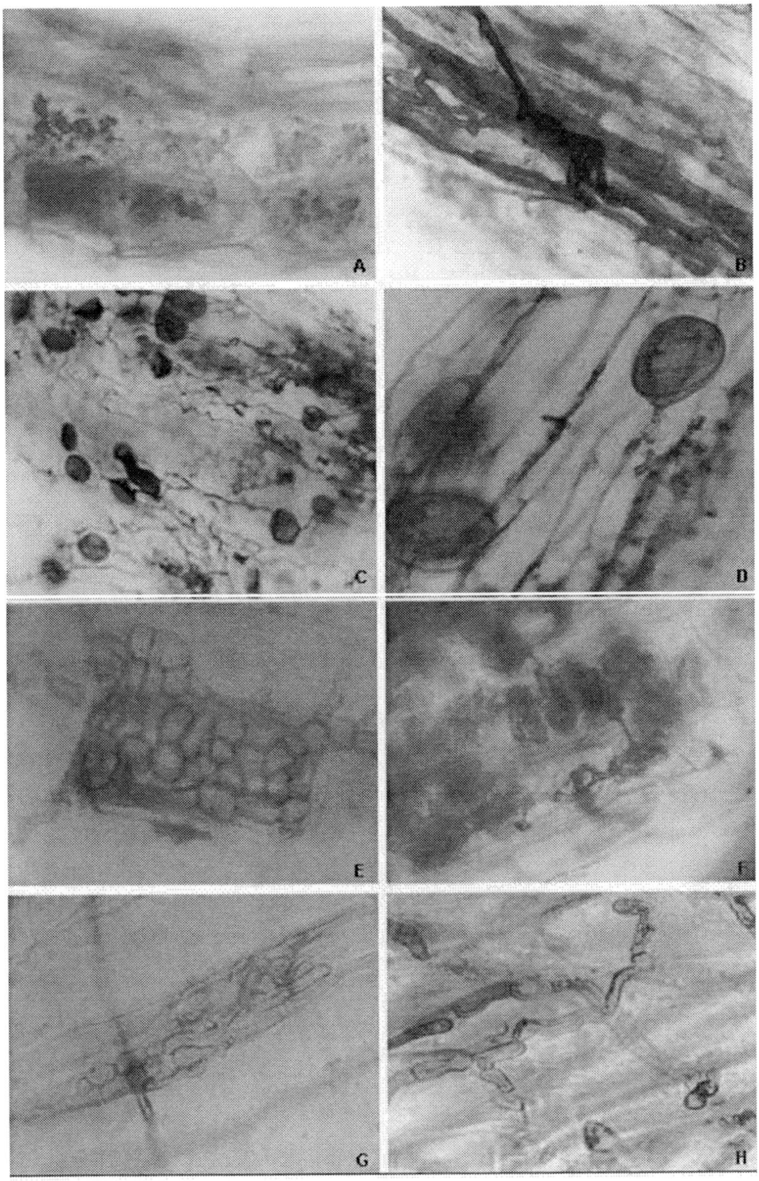

Photos by M. A. Lugo.

Figure 5. AM and DSE colonization in species of the semiarid ecosystems of Argentina. Intraradical structure in roots of Puna grasses. A. Arbuscules. B. Hyphae and coils. C, D. Vesicles.E. DSE. AM and DSE colonization in roots of Gentianaceae F, Arbuscules; G. Hyphae and microesclerotia. H. DSE colonization in roots of *Tillandsia bryoides*(Bromeliaceae).

referred to polymer utilization and exploitation of complex C molecules as cellulose, starch, fatty acid esters, protein, and RNA have been related to their saprotrophic habit. Furthermore, the saprotrophic behavior of DSE is coupled to the facilitation and mobilization of these recalcitrant nutrients from the soil to the host, thus indirectly providing a benefit to the plant (Smith and Read 2008 and references therein). Because of these considerations, and together with their potential from beneficial to injurious for the host, DSE are considered biotrophic facultative (Jumpponen and Trappe 1998) rather than mycorrhizal (Smith and Read *op.cit.*).

In dry environments plant roots are also frequently associated with DSE, a group of dematiaceous anamorphic Ascomycetes that colonize root tissues intra- and intercellularly, forming symbiotic associations that range from mutualism to parasitism (Jumpponen 2001). These fungal symbionts are associated with Dicots and Monocots angiosperms, gymnosperms, ferns and lycopods and are often found in plants that are also associated with AMF.

The DSE fungi included pigmented and phialidic hyphomycetes as *Phialocephala fortii*, *Chloridium pauciflorum*, *Cadophora* (=*Phialophora*) *finlandia* and *Leptodontidium orchidicola*. Molecular analyses of these fungi revealed their affiliation to Helotiales, an order of ascomycetes, mainly as members of Dematiaceae (Smith and Read 2008 and references therein).

DSE are mostly pigmented and grow in cortical parenchyma and into vessel elements of the central vascular cylinder (Barrow and Aaltonen 2001, Barrow 2003). Furthermore, DSE present tight folded hyphae called microesclerotia (Jumpponen and Trappe 1998). Besides, these fungi can generate hyaline hyphae, not melanized, with lipid bodies which are called systemic endophytic fungi or SEF (Barrow 2003) and these colourless hyphae would serve as potential areas for transfer of carbon (Barrow and Aaltonen 2001). Based on morpho-anatomical features and experimental evidence, it has been proposed that these fungi may enhance the performance of plants in arid environments. In particular, it has been suggested that they may enhance nutrient uptake and maintain adequate water relationships for plant survival (Jumpponen 2001, Barrow 2003).

Few studies (Table 1, 3, 5) have been undertaken in South America taking DSE into account (Menoyo et al. 2007, Urcelay 2002, Urcelay et al. 2005, Diehl and Fontenla 2010), and still fewer in the arid and semi-arid environments (Lugo et al. 2009, Salvarredi et al. 2010). In Brazil, for example, only one published report on root DSE from the Atlantic Rain Forest (Grippa et al. 2007) can be found. Nevertheless, the number of reports has been rising lately (Figure 4).

Previous research by Lugo et al. (unpublished data) in the Puna, Argentina, to study root symbioses were based on five sample sites chosen along an altitudinal gradient (3320, 3370, 3520, 3700, 3870 m). The sample sites were located at 22° 51' 35.2'' S, 65° 13' 43.71'' W; 23° 00' 06.8'' S, 65° 22' 07.3'' W; 22° 52' 14'' S, 65° 14' 25.51'' W; 22° 53' 34.1'' S, 65° 16' 0.06'' W; 22° 53' 23.3'' S, 65° 14' 56.1'' W in Jujuy and Salta provinces (Argentina). The physicochemical soil features along the studied gradient were texture sand-gravel-silty to gravel-sandy, 0.745-1.564 percentage of hygroscopic humidity, 0.22-1.15 percentage of oxidale carbon, 0.4-2 percentage of organic matter, 0.1-0.35 of free carbonates, 5.9-7.1 pH, 0.275-1.081 mS/cm of electrical conductivity of extract at 1:2.5 water dilution (at 25 °C), 1.11-6.85 ppm of phosphorous, and each site properties were detailed in Lugo et al. (2005) and Strasser et al. (2005).

In the study by Lugo et al. (unpublished data) twenty native grasses species were analyzed, 10 species presented C_3 photosynthetic pathway, namely: *Bromus catharticum, Danthonia annableae, Danthonia boliviensis, Festuca humilior, Hordeum muticum, Jarava plumosula, Koeleria praeandina, Poa laetevirens, Polypogon interruptus, Vulpia myurus* var. *megaleura*, and 10 were C_4: *Aristida adscensionis, Aristida adscensionis* var. *bromoides, Aristida asplundii, Bouteloua barbata, Bouteloua simplex, Cynodon dactylon* var. *biflorus, Eragrostis nigricans*var. *punensis, Eragrostis mexicana* subsp. *virescens, Microchloa indica*, and *Muhlenbergia rigida*. The grasses' life forms (10 annual and 10 perennial species) were distributed among C_3 and C_4 photosynthetic pathways along five altitudes.

The roots were washed and set in FAA for their subsequent clarification and staining (Grace and Stribley 1991). The finest roots of the 5 individuals of each species were cut into approximately 1 cm segments, mixed, and randomly selected to be mounted in 4 to 5 slides in PVA. For each preparation, they were quantified under the microscope (McGonigle et al. 1990) at 50 X, 100 intersections, determining the arbuscular percentage (% A), hyphas (% H), vesicles (% V) and total root length colonization (% RC) as the % A, % H and % V, and dark septate hyphae (% DSE).

The variables analyzed were percentage of root colonization of AMF (% RC, % A, % V) and DSE (% DSE). These variables were analyzed by Spearman correlations performed considering metabolic type (C_3, C_4), habit (annual, perennial) and height of the study sites (3320, 3370, 3520, 3700, 3870 m). We assessed the effect of altitudinal gradient on the percentage of root colonized by AMF and DSE, and the relationship between two variables (%

RC / % DSE) by Kruskal-Wallis test and post test for multiple comparisons. All analyses were carried out with $\alpha = 0.05$.

All grasses studied were colonized by AMF, 3 grasses presenting exclusively AM (*Cynodon dactylon* var. *biflorus*, *Microchloa indica* and *Muhlenbergia rigida*) and 17 grass species being also associated to DSE (Table 5, Figure 5). Arbuscular mycorrhizal were always found in the roots of Puna grasses; its mean root colonization values were low, but maximal colonization arose to 68 % (Table 4). However, DSE mean root colonization showed lower values than AM mean root colonization (data not shown).

With regard to altitude, AM colonization was found along all different elevations instead of DSE, which were absent in roots of *M. indica* and *M. rigida* at 3320 m, in *Cynodon dactylon* var. *biflorus* at 3370 m and at 3700 m. Mean root colonization of C_3 grasses was lower than C_4 in both association types (AM and DSE). Furthermore, AM and DSE colonization in annual grasses was higher than in perennials grasses. Arbuscular mycorrhizal colonization was slightly correlated with DSE colonization ($r_s = 0.2562$, $p = 0.0026$); when considering AM and DSE colonization, and their relationship with AM structures colonization, there was a slight correlation between % A and % DSE ($r_s = 0.27527$, $P = 0.00122$), while % V was not correlated with % DSE ($r_s = 0.08$, $p = 0.33$).

Moreover, AM colonization differed significantly with altitude ($H = 20.77$, $p = 0.0003$), but this difference could be due to a site effect, because only at one site (3700 m) the highest AM colonization values were reported. Unlike the AM colonization, DSE colonization did not differ between the study sites ($H = 3$, $p = 0.55$). When both variables were considered jointly (RC / DSE), there were significant differences among study sites ($H = 11.29$, $p = 0.0234$) and such differences, as for AM colonization, could be due to a specific effect of sites at 3320 and 3700 m where the ratio between the variables is higher than in the rest of the sites. Correlations between colonization of both symbionts considering each altitude were also conducted. AMF and DSE correlation varied by study site, significant correlations having been found on sites located at 3370 m ($r_s = 0.63172$, $p= 0.00128$), modest at 3700 m ($r_s = 0.48346$, $p= 0.00423$) and a strong correlation at 3870 m ($r_s = 0.76912$, $p= 0.00289$) (Lugo et al. 2007).

When the habit and metabolic type were considered, only in C_4 perennial grasses and AM colonization were correlated with the colonization of DSE ($r_s = 0.53789$, $p= 0.00002$ and $r_s = 0.3369$, $p = 0.0023$, respectively). In conclusion, both symbionts (AMF and DSE) would have a facilitator effect of colonization and C_4 perennial grasses; this could be the case because they are hosts required for a period of time, allowing the establishment of these relationships.

Strikingly and contrary to the expectations, DSE/AM were not negatively correlated. Although DSE are facultative biotrophic fungi with saprotophic capabilities unlike AMF, which are obligatory biotrophs, and both symbiota can be in competition for nutrient capture from the host, each one did not have a detrimental effect upon the other one. Therefore, both symbionts may be functioning synergically, bringing benefits to their host in harsh environments such as the Puna. Furthermore, saprophitism of DSE allows their axenic culture, and this feature could be relevant to use DSE as inoculants in restoration processes. DSEs' skill to grow in culture could signify great advantages on AMF in restoration programs, because AMF obligated biotrophism demands the application of trap culture methodology that is cumbersome and time consuming.

Ultimately, understanding the interactions of DSE and AMF with plants will help us tounderstand plant community structure and function in order to evaluate the consequences of global changes.

Native Soil Microbiota

The plant species listed in the present study are potentially of great importance to the ecology of the semiarid regions. For example, some plant species from the Brazilian States Minas Gerais (Pagano et al. 2010) and Ceará (Pagano, unpublished) showed differences in root colonization among tree species, some of them being highly colonized by AM (Table 2, 3). Shrubs, palm and grasses from the arid and semiarid region of Argentina (Table 4, 5) also showed AM colonization. Since this chapter is primarily concerned with AMF occurrence in Brazilian (Figure 2) and Argentinean arid regions (Figure 3), we have refrained from discussing most studies in dry forests from other countries.

In the Braziian State of Ceará, as expected, AMF are a common and important component in semiarid environments, therefore being important suppliers of relevant information for restoration programs. AM fungal structures were observed in most plant species; however, the colonization pattern and rate varied among them. Vesicles and aseptate hyphae were the most frequent structures present in the plants studied (Pagano and Araújo 2010). AM fungi belonging to *Acaulospora, Ambispora, Entrophospora,Gigaspora, Glomus* and *Scutellospora* were found in rhizospheric soils from trees (Pagano, unpublished), which is in agreement with the fact that they have different life strategies (de Souza et al. 2005), indicating that AMF have different functions in these vegetational types.

Regarding the successful growth of native plants, an understanding of the dependency of these plants on mycorrhizal fungi is urgently needed. At greenhouse, Pagano et al. (2007) showed that phosphorus fertilization enhances the positive effects of AMF, resulting in more vigorous seedlings from dry forest and suggesting that in low fertility soils native trees (*A. peregrina, E. contortisiliquum* and *P. reticulata*)seedlings should be inoculated with AMF to enhance plant growth.

Finally, we also concluded that AM fungi perform the most important symbiotic interactions, despite the fact that the understanding of basic information about key features of AM fungal biology (Fitter et al. 2004) is an urgent need. In the semiarid regions of Argentina and Brazil there is a great abundance of legumes and herbaceous species that can associate mainly with AM fungi. To achieve higher productivity in arid and semiarid soils, it is necessary to select resistant species as well as the symbionts and their associated microorganisms that may increase the survival, biomass and nutrient content of related plants.

POTENTIAL IMPORTANCE FOR CONSERVATION

On the one hand, it was shown that the cutting of plants for wood and charcoal production is a frequent practice in the largest biome of the semiarid region of Brazil (Quesada et al. 2009) and also in Argentina (Cabido and Zak 1999). Annually, millions of hectares of native vegetation in Caatinga dominiumare cut for wood and charcoal production and extensive cattle raising (Figueirôa et al. 2005). Likewise, in Central Argentina, *Aspidosperma quebracho-blanco* and *Schinopsis lorentizii* forests are reduced to 66.000 ha, that represent less than 20 % of their original distribution area (Cabido and Zak 1999). On the other hand, most studies disregard soil biological properties such as soil microbiota, root symbiosis and interactions between plants, fungi, and the environment, in spite of the fact that they participate in soil aggregation (and thus, soil fertility). Differences in the symbiotic interactions in plant species can indicate the importance of this factor for future conservation plans.

As interactions between species, such as predation, competition, parasitism and mutualisms, have a deep influence on the structure of communities, the loss of a species or a change in its abundance, can have a marked effects on ecological processes throughout fragmented landscapes (Bennett and Saunders 2010).

It is known that invasive introduced plants can dominate a native species by interfering with a necessary symbiont of the native species. For example, Garlic mustard (*Alliaria petiolata*) is a highly invasive species in North America (root exudates of garlic mustard are toxic to AMF) (Callaway et al. 2008). The specialized AMF facilitate phosphorus (P) uptake by plants in many terrestrial ecosystems, where P is scarce (P is among the least naturally available of major nutrients) (Sekercioglu 2010). Thus we need to consider the AM–plant interactions with regard to conservation goals.

Moreover, the use of living elements in mixed-fences (very common structures in rural tropical landscapes of the rural semiarid region in Brazil, widespread used for controlling domestic animal movements, protecting cultivated areas, and defining the borders of private properties) not only serve the farmer's needs, but also aid conservation interests, which demands less harvesting of the native vegetation (Nascimento et al. 2009). The use of species with high sprouting capacities contribute to the conservation of other plants that do not have any re-sprouting capacity, such as *Anadenanthera colubrina*, locally preferred for fence posts, and also harvested for civil construction, fuel wood and charcoal (Figueirôa et al. 2005). In conclusion, there is an urgent need for a continuous and integrated study of plant symbioses in these ecosystems, the potential for regeneration in habitatssubjected to disturbance, and, consequently, the wise management of ecosystem goods and services both in Argentina and Brazil.

Case Studies on the Brazilian Dry Vegetation

The Brazilian dry forest provides several ecosystem services, such as wood, charcoal, fruits, fibers, latex, carnauba based waxes, ornamental plants, medicines. The environmental services are provided through the maintenance of the vegetal and animal biodiversity, their ecological importance being known; however, most farmers have no intention to preserve the native cover vegetation (See Pagano and Araújo 2011). Moreover, the importance of native vegetation for water conservation as well as for carbon sequestration and the ecological succession and forest regeneration in South America has been scarcely quantified (Quesada et al. 2009).

Plants from the Caatinga may play several roles for the local communities, but the use of wood is the most commonly recorded one (Sampaio 2002). The role of seasonality in the use of plant resources conditioned preferred plants, whose availability is not affected by the local seasonal pattern, such as shrub and arboreal plants, used for wood (for several uses) and/or in local medicine (stem bark) (see Araújo et al. 2007).

Subsistence agriculture (maize and beans) is the most important economic activity in this region and water availability is the main restriction for agriculture (Antonino et al. 2000). Additionally, there is a great diversity of native fructiferous species of higher economic value for the region: *Spondias tuberosa* (umbuzeiro), *Anacardium occidentale* (cashew) and *Eugenia luschnathiana* (pitombeira) (See Pagano and Araújo 2011), most of them associated with AMF (Amorim et al. 2004, Weber et al. 2004).

In the State of Ceará (Figure 1 A and B), agrosilvopastoral systems were designed for use by small farmers in water protection areas as they are based on drought-tolerant multipurpose species (tolerant to low soil fertility and not susceptible to pests and diseases). The main species are exotic, tree-like legumes such as *Prosopis juliflora* and *Leucaena leucocephala* and the cactaceous *Opuntia ficus-indica*; however, native multipurpose trees that provide a secondary source of income, but serve first of all for the production of wood, but also of fodder, food, non-wood raw materials, medicine, and honey were included (See Pagano and Araújo 2011).

Studies of AM occurrence in three vegetation types from the Ceará State (Figure 2) showed that roots of trees presented AM colonization, in the following order: Carrasco> Caatinga> Dry forest and that the highest rates of colonization were in the dry period, which can be related to water availability in those ecosystems (Pagano, unpublished).

In Brazil, tropical dry forests extend over 3.21% of its territory, mostly in the semi-arid northeastern region; for a review of the current available information on the social–ecological context of Brazilian tropical dry forests, see Espírito-Santo et al. (2009).

In the Semiarid of Minas Gerais, agroforestry systems are an alternative soil management strategy for agricultural production, economic and social growth. The wood provision for local populations using agroforestry systems of mixed native species and *Eucalyptus* in disturbed sites (Pagano et al. 2008, 2009) was based on the ability of woody legumes to take advantage of their ability to fix nitrogen (N), and to form AM. Moreover, threatened tree species can be included (Pagano et al. 2010), thus increasing conservation goals.

CASE STUDIES ON THE ARGENTINIAN DRY PUNA

The Andes (the tropical Andes) are recognized as a center of endemism (Rodrigues et al. 2004) and have been pointed have been ranked first among the 25 hotspots presented by Myers et al. (2000). They are believed to contain at least 20,000 known endemic plants (many more species remain to be discovered there) and are very important for conservation resource priorities.

The Puna is a South American biogeographic region located between 2000 and 4400 m; it is a plateau that extends from northwest Argentina to Southern Peru. This arid environment is characterized by dry and cold climate throughout the year, wide daily temperature range, low precipitation as snow or hail during the year, with rainfall in summer, and intense solar radiation. The soils of the Puna are shallow and immature, very poor in organic matter, sandy and stony. The Puna presents varied floristic districts, and vegetation is constituted by shrub steppes, low shrubs, and mountain grasslands with few trees and herbs (Cabrera and Willink 1973, Morrone 2001).

The Puna agriculture is restricted to a reduced number of native farmers, whose husbandry is based mostly on the culture of maize and horticultural vegetables. The Puna cattle grazing is extensive, their native people use the natural grasses growing in these inorganic soil for pasture, livestock capacity being therefore low. Furthermore, the Puna vegetation is subject to overgrazing by livestock (camels-"llama", sheep, goats), fire and firewood collection (Morrone 2001).

At the same time, the resurgence of traditional livestock of camels in the Puna area (due to, among other factors, the need for sources of animal protein for the native populations and the possible international marketing of llama meat), have intensified the overexploitation of natural pastures and therefore the erosion of these immature soils. For this market to be possible, it is necessary to achieve organic production conditions (required for this type of product in the international market), maintaining soil fertility for sustaining natural pastures.

In the Argentinean Puna, Lugo et al. (2008) studied rhizospheric AM fungi and rhizobaterias associated to native species of grasses to assess the diversity and relationship between these important soil microorganisms in this hash environments. The sampled area were located in the floristic district known as "Jujeño" or "Puna Seca", and varied in altitude from 2000 to 4400 m above sea level. The five sampled sites were placed at different heights (3.320,3.370, 3.520, 3.700, and 3.870 m; Figure 1 A and C). This area is subject to an intense solar radiation, low relative air humidity (10-15%) and

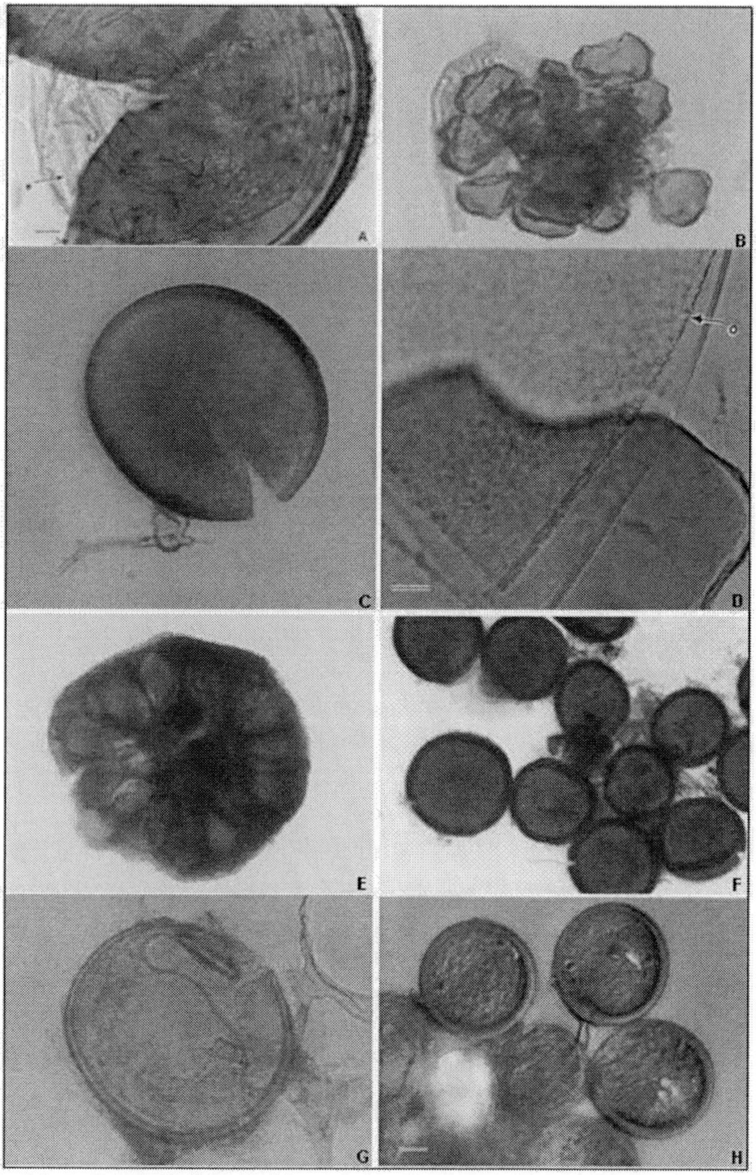

Photos by M. A. Lugo; E-F by E. Menoyo.

Figure 6. Spores of species of AMF found in rhizosphere of grasses from Puna, Argentina. A. *Scutellospora biornata*, detail of the germination shield. B.*Scutellosporabiornata,* auxiliary cells. C. *Gigaspora ramisporophora*. D. *Acaulospora laevis*with ornated (o) inermost wall.E. *Glomus sinuosum*. F. *Glomus ambisporum*.G. *Glomus* sp. H. *Glomus aggregatum*.

large daily thermal differences (16-20°C), resulting in a desertic climatic condition (Martínez Carretero, 1995). Soil from the rhizosphere of fifteen native grasses were analyzed: *Aristida adscensionis*, *Aristida asplundii*, *Bouteloua barbata*, *Bouteloua simplex*, *Bromus catharticus*, *Cynodon hirsutus*, *Danthonia annableae*, *Deyeuxia rigescens*, *Eragrostis nigricans* var. *punensis*,*Jarava subaristata*, *Jarava plumosula*, *Jarava pungens*, *Lycurus setosus*, *Nassella inconspicua*, and *Polypogon interruptus*. These Poaceae presented different photosyntetic pathways (C_3 or C_4) and life forms (annual or perennial); for a detailed explanation of grasses features and distribution at each altitudinal level, see Lugo et al. 2008. The rhizospheric soil from 5 grasses individuals of each species in each sample site were mixed. Each mixed sample was fractioned, one for AMF spores extraction and another for bacterial diversity analysis. The samples were stored at 4°C before processing. AMF spores were obtained from soils by the wet sieving-decanting method (Gerdemann and Nicolson, 1963), followed by centrifugation with 80 % sucrose (Walker et al., 1982). Furthermore, spores quantification was carried out in Petri dishes of 9 cm diameter, provided with a gridline, under dissecting microscope (500 x magnification).

Along the altitudinal gradient ten AMF species were found: *Acaulospora* sp., *A. laevis*, *A. spinosa*, *Gigaspora* sp., *Gi. ramisporophora*,*Glomus* sp., *Gl. aggregatum*, *Gl. ambisporum*, *Gl. sinuosum*, and *Scutellospora biornata* (Figure 6). The highest mean value of spore density was 107.6 spore/100 ml of rhizospheric soil. AMF richness was negatively correlated with the altitude of the studied sites. AMF spore richness was significantly different at each altitude. The highest absolute values for AMF richness (7 species) were found at the lowest altitude (3320 m) and the lowest values were found at the highest altitudes (3700 and 3870 m). Richness did not vary significantly with the metabolic type of grasses.

Spore density of AMF showed negative correlation with altitude. Thus, *Acaulosporaceae*, *Glomeraceae* and *Gigasporaceae* were observed up to 3520 m; *Glomeraceae* was found only at 3700 and 3870 m. Spore density of AMF was not significantly different at each altitude; however, spore densities of *Glomus ambisporum* and *Glomus sinuosum* showed differences with altitude. Furthermore, *Glomus ambisporum* and *Glomus sinuosum* were found along all sites of the gradient. The density of *Glomus ambisporum* decreased with altitude. However, the abundance of *Glomus sinuosum* also decreased with altitude without a clear inverse relationship between abundance and altitude. Specific density of AMF did not change significantly with the plant metabolic type.

Arbuscular mycorrhizal fungal richness and diversity values found in the Puna grasslands were low and similar to those found in arid environments around the world (e. g. Camargo and Dhillon 2002, Cuenca and Lovera 1992, Jacobson 1997) and decreased with altitude. The AMF diversity was lower than in other high altitude grasslands from Argentina (Pérez and Frangi 2001, Lugo and Cabello 2002). Although AMF spore diversity has been related to host species and photosynthetic pathways, no correlation was found between native grasses from Puna and AMF. However, richness and density of AMF were negatively correlated to altitudinal increase. Furthermore, concomitantly to altitude increment and temperature decline, together with the decrease of light energy, could have had an influence on AMF, overlapping the host effect. At the highest altitude, only *Glomeraceae* spores were found as it was also the case in other South American arid environments such as the "Jarillal" or the Namibia desert. Thus, *Glomus* is the predominant genera in extreme environments (Lugo et al. 2008 and references therein).

In this arid and stressed ecosystem, the phylogenetic diversity of culturable and un-culturable rhizospheric bacteria was low (Lugo et al. 2008, Ferrero et al. 2010) when compared to other environments such as salt marshes (Wang et al. 2007), forests (Torvisk et al. 1990) and desert crust soils (Gundlapally and Garcia-Pitchel 2006). Plate counts of aerobic heterotrophic bacteria on R_2A media ranged from $1.31x10^2$ to $8.66x10^4$ CFU g^{-1} of dry weight of rhizospheric soil (Lugo et al. 2008). The apparent heterogeneity of the microbial spatial distribution within the Puna grassland could be explained by specific micro-site features such as temperature, pH, organic matter and plant species (Smalla et al. 2001, Giri et al. 2005, Nunan et al. 2005), rather than the general factors evaluated in this study. Several authors have suggested that plant species have a major selective influence on microbial community structure under controlled conditions (Rovira 1965, Grayston et al. 1998). This is due to differential utilization of C sources taken from the rhizosphere, which may influence the proliferation of particular communities of microorganisms (Grayston et al. 1998, Berg et al. 2002). The functional diversity of microorganisms in different soils may be similar, thereby resulting in a combination of plant and environmental influences on their activity, culturability and proliferation under different conditions (Grayston et al. 1998). The low bacterial diversity and dominance of particular bacterial groups in the native grasses of Puna could be explained by soil type, which may select the most efficient bacteria capable of growth under poor nutrient conditions.

Finally, *Bacillus* and *Pseudomonas* were the dominant bacterial genus in the rhizosphere of the Puna grasses. *Bacillus* have frequently been isolated from arid, semiarid and desert soils, which may be due to its ability to switch to a dormant form under stressful conditions (Dirks 2003). *Bacillus* and *Pseudomonas* genera include nitrate-reducing, free nitrogen-fixing and denitrifying bacteria in soil, and N is the limited nutrient in arid environments because ammonium and nitrate movement from the soil to the plant can be restricted in dry soil and under low temperatures (Dhillon and Zak 1993, Philippot and Germon 2005). It is well known that *Bacillus* function as mycorrhiza helper bacteria (MHB), which stimulate mycelial growth of mycorrhizal fungi or enhance mycorrhizal formation (Garbaye 1994). Additionally, AMF interactions with MHB have synergistic beneficial effects on plant growth that occur through either nutrient solubilization, N fixation by nitrogen-fixing bacteria symbionts or the production of plant hormones (Azcón et al. 1978). Therefore, *Bacillus* and *Pseudomonas* may play an important role in nutrition and the survival of native grasses in the Puna as well as its conservation.

Overgrazing by domestic livestock (goat, sheep, cow and llama) is the most common disturbance that affects ecosystem functioning in the Puna region of Argentina. That is actually a widespread phenomenon throughout all the Argentinean arid and semiarid regions like Chaco and Monte (Figure 3). While restoration of any disturbed area is a complex issue, there exist additional limitations to restoration of arid and semiarid ecosystems given by the low productivity that result in a low ecological resilience of these sites (Aguilera et al. 1999).

Loss of plant biomass is the first visible symptom of overgrazing and thus management and restoration practices generally focus on plant communities, but not on their relationships with soil microorganisms. However, overgrazing causes disturbance of plant-microbe symbioses which are a critical ecological factor in helping further plant reestablishment in degraded ecosystems. Loss of plant biomass by overgrazing is often accompanied by loss of key microorganisms and biological soil properties (Jasper et al. 1991, Schreiner et al. 1997, Schnitzer et al. 2011). These properties largely determine soil quality and fertility, and thus the restoration capacity of the ecosystem. Furthermore, the benefits received by plants from AMF -such as improved nutrient status and increased drought stress tolerance- have the potential to alter the outcome of interspecific plant competition, and therefore, to influence plant succession in natural communities (Allen and Allen 1990, Gange et al. 1993, Hartnett et al. 1993, Smith et al. 1998, Steinaker and Wilson 2008).

The role of mycorrhizae as symbionts in plant succession is relatively well documented (Allen 1989, Allen et al. 1997, Azcón and Barea 1996, Heneghan et al. 2008). Mycorrhizas likely prevent the stagnation of community development, and consequently might play an important role in ecological restoration (Miller and Jastrow 1992, Roldán and Albaladejo 1994, Renker et al. 2004). In the North American sagebrush steppe, for example, AM play an important role in succession and therefore could be used to enhance vegetation recovery to the detriment of a nonmycotrophic invasive species (Renker et al. 2004). Thus, the introduction of indigenous mycorrhizae may be a successful biotechnological tool to aid the recovery of degraded ecosystems.

In semiarid the Mediterranean shrublands of Almeria, Spain, two long-term experiments showed that inoculation with indigenous AMF and with rhizobial nitrogen-fixing bacteria not only enhanced the establishment of key plant species but also increased soil fertility and quality (Requena et al. 1997). Smith et al. (1998) and White et al. (2008) found that AMF inoculation effectively increased AM colonization in a roadside prairie restoration site, and that plots so inoculated had greater cover by native prairie species than uninoculated plots. Restoration of endangered plant species was also considered using AMF in warm temperate ecosystems of North America (Jayachandran and Fisher 2008). In all these cases, a straightforward manipulation of single biological components of the system can be useful in the restoration of a site. These single-factor manipulations may in fact produce cascading effects on several ecosystem attributes and can result in unintended recovery trajectories (Heneghan et al. 2008).

It is concluded that in arid and semiarid ecosystems of Argentina such as the Puna region, many native shrub and grass species are highly colonized by AMF (Tables 4, 5). There are few studies, however, evaluating the effect of mycorrhizal amendment at restoration sites and mycorrhizal dependence on native plants. Thus, it becomes necessary to multiply the efforts in examining and evaluating the effectiveness of the introduction of target native species of plants associated with a managed community of microbial symbionts as a potential technology to recover degraded areas in semiarid and arid regions of Argentina.

Conclusions

On the one hand, it was shown that research on native species of both plant (also for endangered vegetation) and fungi has received more attention for ecosystem restoration, for mitigation of the expected effects on biodiversity and to deal with climate changes both in Argentina and Brazil.

On the other hand, despite adequate studies on the mycotrophy of plants are fundamental to improve their conservation, few reports in South America pointed to the importance of the soil biological processes in influencing biodiversity conservation of natural ecosystems. We underlined the importance of DSE in roots of plants, as possible mutualistic as well as the study of plant symbioses in order to improve the conservation of the biomes mentioned. Similarly, the rhizospheric bacterial communities associated to native grasses contribute greatly to this conservation, because they enhance the inorganic nutrient movement from the soil to the plant and also mycorrhizal formation in arid environments. AMF interactions with MHB have synergistic beneficial effects on plant growth that occur through either nutrient solubilization, N fixation by nitrogen-fixing bacteria symbionts or the production of plant hormones.

The strong use pressure (most for wood uses) will need to be evaluated for management and conservation purposes, and efforts must be able to count on the cooperation of a multidisciplinary group, and must take into account the symbioses and root interactions of the plant species. Moreover, since this chapter is primarily concerned with native plants, we have refrained from discussing economic plants that most studies have often focused on.

Our review also indicates the need for studies that consider the ecological processes and the use of native plants by people together, as well as the need to increase studies on the Brazilian Dry forest and the Andean Puna.

Research on native species of both plant and fungi has received more attention for ecosystem restoration, and for mitigation of the expected effects of land use change on biodiversity. In spite of the fact that adequate researches on the mycotrophy of plants are fundamental to improve their conservation, few studies in South America pointed to the importance of the soil biological processes in influencing biodiversity conservation of natural ecosystems. We have attempted to highlight the importance of DSE in roots of plants as possible mutualistic, as well as the study of plant symbioses in order to improve the conservation of these biomes. The introduction of mycorrhizae associated to the management of key native species of plants is a technology to be developed in order to recover degraded sites at the Caatinga dominiumand Puna regions. The strong local pressure of land use (mainly for forage and wood uses) should be evaluated for management and conservation purposes, even though all products of these uses are directed towards subsistence, with little contribution to commerce.

Efforts directed to resource management and conservation must be able to count on the cooperation of multidisciplinary groups, and must take into account the symbioses and root interactions of the plant species. Moreover, special attention should be given to some species such as the ones referred to as threathened in the Red List, but also those endemic or native ones, which are being "persecuted" by the advance of agricultural frontiers.

Acknowledgments

Dr Marcela Pagano is grateful to the Council for the Development of Higher Education at Graduate Level, Brazil (CAPES), Process PRODOC 2125/2008, for a Post-doctoral scholarship granted. Drs Marcela A. Ferreroand Mónica A. Lugo are Researchers from Consejo Nacional de Investigaciones Científicas y Tecnológicas (CONICET), Argentina. Biologist Eugenia Menoyo has a fellowship from CONICET, Argentina.

References

Aguilera, MO; Demaría, MR; AO; Avila,AO; Steinaker, DF. Impacto de la intensificación del uso de la tierra en la diversidad vegetal: un estudio en los pastizales semiáridos de San Luis, Argentina. In: Biodiversidad y uso de la tierra. Conceptos y ejemplos de America Latina. Matteucci, SD;Solbrig,OT;Morello,J. y Halffter, G, editors,EUDEBA, Buenos Aires, Argentina; 1999; pp 515–528(In Spanish).

Allen, EB. The restoration of disturbed arid landscapes with special reference to mycorrhizal fungi. *J. Arid Environ.*, 1989,17, 279–286.

Allen, EB; Allen, MF. The mediation of competition by mycorrhizae insuccessional and patchy environments. In: Perspectives on plant competition. Grace, JB; Tilman, D., editors,Academic Press, Inc., San Diego; 1990; pp. 367–389.

Allen, EB; Esquejel, I; Sigüenza, C. Role of mycorrhiza in restoration of marginal and derelict land and ecosystem sustainability. In: Palm, ME; Chapela, IH, editors. *Mycology in Sustainable Development: Expanding concepts, vanishing borders*. Parkway Publishers, Inc. Boone, North Carolina; 1997.

Almeida, FAG; Martins Junior, W; Almeida, FCG. Fenologia comparativa de dois clones enxertados de cajueiro anão em condições de irrigação. *Ciência Rural*, 2002, 32, 221–228 (In Portuguese).

Amorim, SMC; Paim, ACB; Silva, MG. Estudo ecofisiológico sobre endomicorrizas. *Revista Biotecnologia Ciência and Desenvolvimento*, 2004, 33, 23–26 (In Portuguese).

Antonino, ACD; Sampaio, EVSB; Dall'Olio, A; Salcedo IH. Balanço hídrico em solo com cultivos desubsistência no semi-árido do nordeste do Brasil.*Revista Brasileira de Engenharia Agrícola e Ambiental*, 2000, 4, 29–34 (In Portuguese).

Araújo, FP; Santos, CAF; Calvacante, NB; Rezende, GM. Influência do período de armazenamento das sementes de umbuzeiro na sua germinação e no desenvolvimento de plântula. *Revista Brasileira de Armazenamento*, 2001, 26, 36–39 (In Portuguese).

Araújo, FS; Martins, FR. Fisionomia e organização da vegetação do carrasco no planalto da Ibiapaba, estado do Ceará. *Acta Botanica Brasílica,* 1999, 13, 1–14 (In Portuguese).

Araújo, FS; Sampaio, EVSB; Figueiredo, MA; Rodal, MJN; Fernandes, AG. Composição florística da vegetação de carrasco, Novo Oriente, CE. *Rev. Bras. Bot.*, 1998, 21, 2, 105–116 (In Portuguese).

Araújo, FS; Costa, RC; Lima, JR; Vasconcelos, SF; Girão, LC; Sousa-Sobrinho, M; Bruno, MMA; Sousa, SSG; Nunes, EP; Figueiredo, MA; Limaverde, LW; Loiola, MIB. Floristics and life-forms along a topographic gradient, central-western Ceará, Brazil.*Rodriguesia*, 2011, 62, 2, 341–366.

Araújo, EL; Castro, CC; Albuquerque, UP. Dynamics of Brazilian Caatinga a Review Concerning the Plants, Environment and People. *Functional Ecosystems and Communities*, 2007, 1, 1–29.Araújo, EL. Diversidade de herbáceas na vegetação da caatinga. In: Jardin EAG, Bastos MNC, Santos JUM, editors. *Desafios da Botânica Brasileirano Novo Milênio: Inventário, Sistematização e Conservação da DiversidadeVegetal.* Belém: Sociedade Brasileira de Botânica; 2003; 82–84 (In Portuguese).

Azcón-Aguilar, C., Barea, J. M. Arbuscular mycorrhizas and biological control of soilborne pathogens – an overview of the mechanisms involved. *Mycorrhiza*, 1996, 6, 457–464.

Azcón, R; Azcón-Aguilar, C; Barea, J.M; Effect of plant hormones present in bacterial cultures on the formation and responses to VA endomycorrhiza.*New Phytologist*, 1978, .80, 359–364.

Baar, J. Restoration of plant communities in The Netherlands through the application of arbuscular mycorrhizal fungi. *Symbiosis*, 2010, 52, 87–94.

Barros, LM; Cavalcanti, JJV; Paiva, JR; Crisóstomo, JR; Corrêa, MPF; Lima, AC. Seleção de clones de cajueiro-anão para o plantio comercial no Estado do Ceará. *Pesquisa Agropecuária Brasileira*, 2000, 35, 2197–2204 (In Portuguese).

Barrow, J.R. Atypical morphology of dark septate fungalroot endophytes of *Bouteloua* in arid southwestern USArangelands. *Mycorrhiza,* 2003,13, 239–247.

Barrow, J.R., Aaltonen, R.E. Evaluation of the internal colonization of *Atriplex canescens* (Pursh) Nutt. Roots by dark septate fungi and the influence of host physiological activity. *Mycorrhiza,* 2001,11, 199–205.

Becerra, A; Cabello, M; Caballero, C; Renison, D; Soteras, F. Infectividad de hongos micorrícico-arbusculares en los bosques de *Polylepis australis* de las Sierras Grandes de Córdoba. In: XXII Congreso Argentino de la Ciencia del Suelo, Rosario, Argentina; Libro de resúmenes, 2010, p. 87 (In Spanish).

Bennett, AF; Saunders, DA. Habitat fragmentation and landscape change. In: Sodhi, N.S.; Ehrlich, P.R. *Conservation Biology for All*; 2010; pp. 88–104.

Berg, G; Roskot, N; Steidle, A; Eberl, L; Zock, A; Smalla, K. Plant-dependent genotypic and phenotypic diversity of antagonistic rhizobacteria isolated from different *Verticillium* host plant. *App. Environ. Microbiol.,* 2002, 68, 3328–3338.

Berg, G; Smalla K. Plant species and soil type cooperatively shape the structure and function of microbial communities in the rhizosphere. *FEMS Microbiol. Ecol,* 2009, 68, 1–13.

Bradshaw, AD. Introduction and philosophy. In: *Handbook of ecological restoration,* Perrow, M.R.; Davy, A.J., editors. Cambridge University Press, Cambridge; 2002; pp. 1–9.

Cabello, MN. Mycorrhizas and hydrocarbons. In: *Fungi in Bioremediation,* Gadd GM, editor, Cambridge University Press, Cambridge; 2002; pp. 456–471.

Cabido, MR; Zak, MR. Vegetación del Norte de Córdoba, Imprenta Nico, Argentina; 1999 (In Spanish).

Cabrera, AL. Fitogeografía de la República Argentina.*Bol. Soc. Arg. Bot.* 1971, 14, 1–42 (In Spanish).

Cabrera, A.L. Territorios fitogeográficos de la República Argentina, in: Enciclopedia argentina de Agricultura y Jardinería, 2da. Edición, II (1). ACME Ediciones, Buenos Aires, 1976, pp. 1–85 (In Spanish).

Cabrera, AL; Willink, A. *Biogeografía de América Latina*. OEA, Whashington, DC; 1980 (In Spanish).

Callaway, R M; Cipollini, D; Barto, K et al. Novelweapons: invasive plant suppresses fungal mutualismsin America but not in its native Europe. *Ecology*, 2008, 89, 1043–1055.

Camargo-Ricalde, SL; Dhillion, SS. Endemic *Mimosa* species can serve as mycorrhizal "resource islands" within semiarid communities of the Tehuacán-Cuicatlán Valley, Mexico. *Mycorrhiza*, 2002, 3, 129–136.

Chapin, F. S., III, Zavaleta, E. S., Eviner, V. T., Naylor, R. L., Vitousek, P. M., Reynolds, H. L., et al. Consequences of changing biodiversity. *Nature*, 2000, 405, 234–242.

Chaudhry, MS; Batoo, Z; Khan AG. Preliminary assessment of plant community structure and arbuscular mycorrhizas in rangeland habitats of Cholistan desert, Pakistan. *Mycorrhiza*, 2005, 15, 606–611.

Chowdhury, S.P., Schmid, M., Hartman , A., Tripathi, A.K.Diversity of 16S-Rrna and *nif*H genes derived from rhizosphere soil and roots of an endemic drought tolerant grass, *Lasiurus sindicus*. *European Journal of Soil Biology*, 2009, 45, 114–122.

Costa, RC; Araújo, FS; Lima-Verde, LW. Flora and life-form spectrum in an area of deciduous thorn woodland (caatinga) in northeastern, Brazil. *Journal of Arid Environments*, 2007, 68, 237–247.

Cuenca, G; Andrade, Z; Escalante, G. Diversity of Glomalean spores from natural disturbed and revegetated communities growing on nutrient-poor tropical soils. *Soil Biol. Biochem.*, 1998, 30, 711–719.

Cuenca, G; Lovera, M, Vesicular-arbuscular mycorrhizae in disturbed and revegeted sites from La Gran Sabana, Venezuela. *Canadian Journal Botany*, 1992, 70,73–79.

Díaz, S; Fargione, J; Chapin, FS; Tilman, D. Biodiversity loss threatens human well-being. *PLoS Biology*, 2006, 4, 1300–1305.

Diehl, P; Fontenla, S. Arbuscular mycorrhizal infection in two morphological root types of *Araucaria araucana* (Molina) K. Koch. *Revista Argentina Microbiología*, 2010, 42, 133–137.

Duelli, P; Obrist, MK. Biodiversity indicators: the choice of values and measures. *Agriculture, Ecosystems and Environment*, 2003, 98, 87–98.

Dhillon, S.S., Zak, J.C., 1993. Microbial dynamics in arid ecosystem: desertification and the potential role of mycorrhizas. *Revista Chilena de Historia Natural*, 66, 253–270.

Dirks, A. 2003. The dynamics spore. *Proceeding of the National Academic of Science*, 100, 3007–3009.

Espírito-Santo, MM; Sevilha, AC; Anaya, FC; Barbosa, R; Wilson Fernandes, G; Sanchez-Azofeifa G. A, et al. Sustainability of tropical dry forests: Two case studies in southeastern and central Brazil. *Forest Ecology and Management,* 2009, 258, 922–930.

Ferrero, MA; Menoyo, E; Negritto, MA; Lugo, MA; Farias, ME; Anton, AM; Siñeriz, F. Molecular characterization and in situ detection of bacterial communities associated to rhizosphere soil of high altitude native *Poaceae* from Andean Puna region. *Journal of Arid Environment*, 2010, 74, 1177–1185.

Figueirôa, JM; Perein, FGC; Drumond, M; Araújo, EL. Madeireiras. In: Sampaio, EVSB; Perein, FGC; Figueirôa, JM; Santos-JR, AGS, editors. *Espécies da flora Nordestina de importância econômica potencial.* Recife: Associação Plantas do Nordeste; 2005; pp. 101–133 (In Portuguese).

Fitter, A. H., and R. S. R. Fitter. Rapid changes in flowering time in British plants. *Science,* 2002, 296, 1689–1691.

Fitter, AH; Heinemeyer, A.; Husband, R; Olsen, E; Ridgway, KP; Staddon, PL. Global environmental change and the biology of arbuscular mycorrhizas: gaps and challenges. *Can. J. Bot.* 2004, 82, 1133–1139.

Fontenla, S; Havrylenko, M; Rosso, P. Micorrizas vesículo-arbusculares en *Astrocedrus chilensis.Suelo y Planta*, 1991, 1, 251–255 (In Spanish).

Fontenla, S; Godoy, R; Rosso, P; Havrylenko, M. Root associations in *Astrocedrus* forest and seasonal dynamics of arbuscular mycorrhizas.*Mycorrhiza*, 1998, 8, 29–33.

Fontenla, S; Puntieri, J; Ocampo, JA. Mycorrhizal associations in the Patagonian steppe, Argentina. *Plant and Soil*, 2001, 233, 13–29.

Fracchia, S; Aranda, A; Gopar, A; Silvani, V; Fernandez, L; Godeas, A. Mycorrhizal status of plant species in the Chaco Serrano Woodland from central Argentina. *Mycorrhiza*, 2009, 19, 205–214.

Fuchs, B; Haselwandter K. Red list plants: colonization by arbuscular mycorrhizal fungi and dark septate endophytes. *Mycorrhiza*, 2004, 14, 277–281.

Fuchs, B;Haselwandter, K. Arbuscular Mycorrhiza of Endangered PlantSpecies: Potential Impacts on RestorationStrategies. In:*Mycorrhiza, State of the Art, Genetics and Molecular Biology, Eco-Function, Biotechnology, Eco-Physiology, Structure and Systematics,* Varma A., editor; 2008; pp. 565–579.

Gange, AC; Brown, VK; Sinclair, GS. Vesicular-arbuscular mycorrhizal fungi: a determinant of plant community structure in early succession. *Functional Ecology*, 1993, 7,616–622.

Garbaye, J. 1994. Helper bacteria: a new dimension to the mycorrhizal symbiosis. *New Phytologist*, 128, 197–210.

Gerdemann, JW; Nicolson, TH. Spores of a mycorrhizal *Endogone species* extracted from the soil by wet sieving and decanting. *Transaction British Mycological Society*, 1963, 46, 235–244.

Giri, B., Giang, P.H., Kumari, R., Prasad, R., Varma, A. Microbial Diversity in Soils, in: Buscot, F., Varma, A., editors.Microorganism in soils: roles in genesis and functions. Springer, Germany; 2005; pp. 19–55.Grace, C., Stribley, D.P. A safer procedure for routine staining of vesicular arbuscular mycorrhizal fungi. *Mycological Research*, 1991, 95, 1160–1162.

Grayston, SJ; Wang, S; Campbell, CD; Edwards,A.C. Selective influence of plant species on microbial diversity in the rhizosphere. *Soil Biology and Biochemistry,* 1998, 30, 3, 369–378.

Grippa, CR; Hoeltgebaum, MP; Stürmer, SL. Occurrence of arbuscular mycorrhizal fungi in bromeliad species from the tropical Atlantic forest biome in Brazil. *Mycorrhiza*, 2007, 17, 3, 235–240.

Grman, E; Suding, K. Within-year soil legacies contribute to strong priority effects of exotics on native California grassland communities. *Restoration Ecology*, 2010, 18, 5, 664–670.

Gundlapally, S.R., Garcia-Pichel, F. The community and phylogenetic diversity of biological soil crust in the Colorado Plateau studied by molecular fingerprinting and intensive cultivation. *Microbial Ecology*, 2006, 52, 354–357.

Gyssels, J; Poesen, J; Bochet, E; Li, Y. Impact of plant roots on the resistance of soils to erosion by water: A review. *Progress in Physical Geography*, 2005, 29, 189–217.

Hartnett, DC; Hetrick, BAD; Wilson, GWT; Gibson, DJ. Mycorrhizal influence of intra- and interspecific neighbour interactions among co-occurring prairie grasses. *Journal of Ecology*, 1993, 81, 787–795.

Hausmann, NT; Hawkes, CV. Order of plant host establishment alters the composition of arbuscular mycorrhizal communities. *Ecology*, 2010, 91(8), 2333–2343.

Hawkes, CV; Belnap, J; D'Antonio, C; Firestone, MK. Arbuscular mycorrhizal assemblages in native plant roots change in the presence of invasive exotic grasses. *Plant and Soil*, 2006, 281, 369–380.

Hawksworth, DL. The magnitude of fungal diversity: the 1.5 million species estimate revisited. *Mycological Research*, 2001, 105, 1422–1432.

Heneghan, L; Miller, SP; Baer, S; Callaham, MA; Montgomery, J; Pavao-Zuckerman, M; Rhoades, CC; Richardson, S. Integrating soil ecological knowledge into restoration management. *Restoration Ecology*, 2008, 16, 608–617.

Herrera, MA; Salamanca, CP; Barea, J. Inoculation of woody legumes with selected arbuscular Mycorrhizal fungi and rizobia to recover desertified Mediterranean ecosystems. *Appl. Environ. Microbiol.*, 1993, 59, 129–133.

IUCN Red List of threatened species. International Union for Conservation of Nature and Natural Resources. http://www.iucnredlist.org/info/stats, 2006.

Jayachandran, K; Fisher, J. Arbuscular mycorrhizae and their role in plant restoration in native ecosystems. In: Siddiqui, Z.; Akhtar, M.; Futai, K. (editors). Media, Springer Netherlands; 2008; pp. 195–209.

Jasper, DA; Abbott, LK; Robson, AD. The effect of soil disturbance on vesicular arbuscular mycorrhizal fungi in soils from different vegetation types. *New Phytologist,* 1991, 118, 471–476.

Jacobson, KM. Moisture and substrate stability determine VA-mycorrhizal fungal community distribution and structure in an arid grassland. *Jounal Arid Environment*, 1997, 35, 59–75.

Johnson, NC. Responses of *Salsola kali* and *Panicum virgatum* to mycorrhizal fungi, phosphorus and soil organic matter: implications for reclamation. *Journal Applied Ecology*, 1998, 35, 86–94.

Jumpponen, A. Dark septate endophytes—are they mycorrhizal? *Mycorrhiza*, 2001, 11, 207–211.

Jumpponen, A; Trappe, JM. Dark septate endophytes: a review of facultative biotrophic root-colonizing fungi. *New Phytol.*, 1998, 140, 295–310.

Kuske, CR; Ticknor, LO; Miller, ME; Dunbar, JM; Davis, JA; Barns, SM; Belnap, J. Comparison of soil bacterial communities in rhizospheres of three plant species and the interspaces in an arid grassland. *Applied and Environmental Microbiology*, 2002, 68, 1854–1863.

Likar, M; Bukovnik, U; Kreft, I; Chrungoo, NK; Regvar, M. Mycorrhizal status and diversity of fungal endophytes in roots of common buckwheat (*Fagopyrum esculentum*) and tartary buckwheat (*F. tataricum*). *Mycorrhiza,* 18, 6-7, 309–315.

Lovera, M; Cuenca, G.Diversity of arbuscular mycorrhizal fungi (AMF) and mycorrhizal potential of the soil from a natural and a disturbed savannah from La Gran Sabana, Venezuela. *Interciencia,* 2007, 32 (2), 108–114 (In Spanish).

Lucena, RFP; Araújo, EL; Albuquerque, UP. Does the Local Availability of Woody *Caatinga* Plants(Northeastern Brazil) Explain Their Use Value? *Economic Botany*, 2007, 61(4), 347–361.

Lugo, MA; Cabello, MN. Native arbuscular mycorrhizal fungi (AMF) from mountain grassland (Córdoba, Argentina) I. Seasonal variation of fungal spore diversity. *Mycologia*, 2002, 94, 579–586.

Lugo, MA; Strasser, E; Perino, E; Fernández, E. Estudio con fines geoquímicos del suelo rizósferico de Poaceae en pastizales de la Puna jujeña. II. Fósforo asimilable. III Congreso Argentino de Química Analítica. Merlo, San Luis; 2005; pp. 8–12 (In Spanish).

Lugo, MA; Anton, AM; Cabello, MN. Arbuscular mycorrhizas in the *Larrea divaricata* scrubland of the arid "Chaco", Central Argentina. *Journal Agriculture Technology,*2005, 1, 163–178.

Lugo, MA; Menoyo, E; Negritto, MA; Anton, AM. Micorrizas arbusculares *versus* "septados oscuros" en pastos puneños. XXXI Jornadas Argentina de Botánica, Corrientes, Argentina; *Boletín Sociedad Argentina Botánica (Suppl.)*, 2007, 42, 210 (In Spanish).

Lugo, MA; Ferrero, MA; Menoyo, E; Estévez, MC; Siñeriz, F; Anton, AM. Arbuscular mycorrhizal fungi and rhizospheric bacteria diversity along an altitudinal gradient in South American Puna grassland. *Microbial Ecology*, 2008, 55, 705–713.

Lugo, MA; Molina, MG; Crespo, EM. Arbuscular mycorrhizas and dark septate endophytes in bromeliads from South American arid environment. *Symbiosis*, 2009, 47, 17–21.

Machado, IS. Biologia e fenologia. In: Sampaio, EVSB; Mayo, SJ; Barbosa,MEV, editors. *Pesquisa Botânica Nordestina: Progresso e Perspectivas.*Recife: Editora Universitária; 1996; 161–177 (In Portuguese).

Maron, PA; Mougel, C; Ranjard, L. Soil microbial diversity: Methodological strategy, spatial overview and functional interest. *C. R. Biologies,* 2011 (in press).

Martínez Carretero, E. La Puna argentina: División y distritos florísticos y demarcación general. Con mapa a escala 1/ 2.500.000. *Boletin de la Sociedad Argentina deBotanica,* 1995, 31,1–2, 27–40 (In Spanish).

May, RM. Why should we be concerned about loss of biodiversity. *C. R. Biologies*, 2011 (in press).

McGuire, KL; Treseder, KK. Microbial communities and their relevance for ecosystem models: Decomposition as a case study. *Soil Biology and Biochemistry*, 2010, 42, 529–535.

Menoyo, E; Becerra, AG; Renison, D. Mycorrhizal associations in *Polylepis* woodlands of Central Argentina. *Canadian Journal Botany*, 2007, 85, 526–531.

Menoyo, E; Renison, D; Becerra, AG. Arbuscular mycorrhizas and performance of *Polylepis australis* trees in relation to livestock density. *Forest Ecology and Management*, 2009, 258, 2676–2682.

Miller, RM; Jastrow, JD. The application of VA mycorrhizae to ecosystem restoration and reclamation. In: *Mycorrhizal functioning: an integrative plant-fungal process*, Allen, MF, editor. Chapman and Hall, New York; 1992; pp. 438–467.

Moorman, T; Reeves, FBJ. The role of endomycorrhiza in vegetation practices in the semi-arid west. II. A bioassay to determine the effect the land disturbance on endomycorrhizal populations. *American Journal Botany*, 1979, 66, 14–18.

Morrone, JJ. *Biogeografía de América Latina y del Caribe*. Manuales y Tesis. SEA, México, 3er vol., 2001 (In Spanish).

Myers, N; Mittermeier, RA; Mittermeier, CG; da Fonseca, GAB; Kent, J. Biodiversity hotspots for conservation priorities. *Nature*, 2000, 403, 853–858.

Nascimento, VT; Sousa, L.G; Alves, AGC; Araújo, EL; Albuquerque, UP. Rural fences in agricultural landscapes and their conservation role in an area of caatinga (dryland vegetation) in Northeast Brazil. *Environ. Dev. Sustain.*, 2009, 11, 1005–1029.

Newsham, KK; Upson, R; Read, DJ. Mycorrhizas and dark septate root endophytes in polar regions. *Fungal Ecology*, 2009, 2, 10–20.

Newsham, KK. Meta-analysis of plant responses to dark septate root endophytes. *New Phytologist*, 2011, 190, 783–793.

Nunan, N; Daniell, TJ; Singh, BK; Papert, A; Mc Nicol, JW; Prosser, JI. Links between plant and rhizoplane bacteria communities in grassland soils, characterized using molecular techniques. *Applied and Environmental Microbiology*, 2005, 71, 6784–6792.

Oehl, F; Laczko, E; Bogenrieder, A; Stahr, K; Bösch, R; van der Heijden,M; Sieverding, E. Soil type and land use intensity determine the composition of arbuscular mycorrhizal fungal communities. *Soil Biology and Biochemistry,* 2010, 42, 5, 724–738.

Pagano, MC; Cabello, MN; Scotti, MR. Phosphorus response of three native Brazilian trees to inoculation with four arbuscular mycorrhizal fungi. *J. Agric. Technol.* 2007, 3: 231–240.

Pagano, MC; Cabello, MN; Bellote, AF; Sa, NM; Scotti, MR. Intercropping system of tropical leguminous species and *Eucalyptus camaldulensis*, inoculated with rhizobia and/or mycorrhizal fungi in semiarid Brazil. *Agrofor. Syst.*, 2008, 74, 231–242.

Pagano, MC; Scotti, MR; Cabello, MN. Effect of the inoculation and distribution of mycorrhizae in *Plathymenia reticulata* Benth under monoculture and mixed plantation in Brazil. *New Forests*, 2009, 38,197–214.

Pagano, MC; Cabello, MN; Scotti, MR. Agroforestry in dry forest, Brazil: Mycorrhizal Fungi Potential. In: LR Kellymore, editor. *Handbook on Agroforestry: Management Practices and Environmental Impact.* Nova Science Publishers, New York; 2010; pp. 367–387.

Pagano, MC; Araújo FS.Semiarid vegetation in Brazil: Biodiversity, Impacts and Management. In: Degenovine KM, editor. *Semi-Arid Environments: Agriculture, Water Supply and Vegetation.* Nova Science Publishers, New York; 2011; pp. 99–114.

Pagano, MC; Utida, MK; Gomes, EA; Marriel, IE; Cabello, MN; Scotti, MR. Plant-type dependent changes in arbuscular mycorrhizal communities as soil quality indicator in semi-arid Brazil. *Ecological Indicators*, 2011, 11 (2), 643–650.

Pagano, MC; Cabello MN. Mycorrhizas in natural and restored riparian zones. In: Pagano MC, editor Mycorrhiza: Occurrence and Role in Natural and Restored Environments. Nova Science Publishers, Hauppauge, NY, 2011. ISBN: 978-1-61209-226-3.

Palm, ME; Chapela, IH. *Mycology in Sustainable Development: Expanding concepts, vanishing borders*. Parkway Publishers, Inc. Boone, North Carolina; 1997.

Pardo, OC; Tomasini, D; Maccagno, P; Quadri, V; Panigatti, JL; Montes, L; Collado, A; Maceira, N; Abrahan, E; Karlin, U; Braun Wilke, R; Merega, JL; Stahringer, G; Gonzales, ME; Paris, S. Segundo informe nacional para la implementación de la C de las NU de lucha contra la desertificación, 2002 (In Spanish).

Patreze, CM; Cordeiro L. Nitrogen-fixing and vesicular-arbuscular mycorrhizal symbioses in some tropical legume trees of tribo Mimoseae. *Forest Ecol. Manag.*, 2004, 196, 275–285.

Philippot, L; Germon, JC. Contribution of bacteria to initial input and cycling of nitrogen in soils, in: Buscot, F., Varma, A. (Eds.), Microorganism in soils: roles in genesis and functions, Springer, Germany; 2005; pp. 159–191.

Prado, DE. As Caatingas de América do Sul. In: Leal, IR; Tabarelli, M; Silva, JMC; editors. *Ecologia e Conservação da Caatinga*. Recife: Ed. Universitária da UFPE; 2003; pp. 1–74 (In Portuguese).

Quesada, M; Sanchez-Azofeifa, GA; Alvarez-Añorve, M; Stoner, KE et al. Succession and management of tropical dry forests in the Americas: Review and new perspectives. *Forest Ecology and Management*, 2009, 258, 1014–1024.

Quijas, S; Schmid, B; Balvanera, P. Plant diversity enhances provision of ecosystem services: A new synthesis.*Basic and Applied Ecology*, 2010, 11, 582–593.

Reeves, FB; Wagner, D; Moorman, T; Kiel, J. The role of endomycorrhiza in vegetation practices in the semi-arid west. I. A comparison of incidence of mycorrhizal in severely disturbed vs. natural environments. *American Journal Botany*, 1979, 66, 6–13.

Requena, N., Jimenez, I., Toro, M., Barea, J.M., 1997. Interactions betweenplant-growth-promoting rhizobacteria (PGPR), arbuscular mycorrhizal fungi and *Rhizobium* spp. in the rhizosphere of Anthyllis cytisoides, a model legume for revegetation in Mediterranean semi-arid ecosystems, *New Phytol*., 136, 667–677.

Renker, C; Zobel, M; Opik, M; Allen, M; Allen, E; Vosátka, M. Structure, dynamics, and restoration Structure, dynamics, and restoration of plant communities: do arbuscular mycorrhizae matter? In: *Assembly Rules and Restoration Ecology Bridging the Gap Between Theory and Practice*; 2004.

Rizzini, CT. *Tratado de fitogeografia do Brasil: aspectos ecológicos, sociológicos e florísticos*. 2nd Edition. Rio de Janeiro: Âmbito Cultural Edições Ltda.; 1997 (In Portuguese).

Rodrigues, ASL; Andelman, SJ; Bakarr, MI et al. Effectiveness of the global protected area network in representing species diversity, *Nature*, 2004, 428, 640–643.

Roldán A, Albaladejo, J. Effect of mycorrhizal inoculation and soil restoration on the growth of *Pinus halepensis* seedlings in a semiarid soil. *Biology and Fertility of Soils*, 1994, 18, 2, 1430–149.

Rovira AD.Interactions between plant roots and soil micro-organisms.*Annu Rev Microbiol*1965, 19: 241–266.

Ruiz-Jaen, MC; Aide, TM. Restoration success: How is it being measured? *Restoration Ecology*, 2005, 13, 569–577.

Ruthsatz, B.*Pflnzengesellschaften und ihre Lebensbedingungrn in den Andinen Halbwiisten Nordwest Argentiniens.* Dissertactiones Botanicae. Band 39. J. Cramer Fl 9490, Vaduz; 1977; pp. 1–168 (In German).

Salvarredi, LA; Crespo, EM; Menoyo, E; Filippa, E; Barboza, G; Lugo, MA... Micorrizas arbusculares y endófitos septados oscuros en *Gentianaceae* nativas de la Argentina. *Boletín Sociedad Argentina Botánica,* 2010,45, 223–229 (In Spanish).

Sampaio, EVSB. Uso das plantas da caatinga. In: Sampaio, EVSB; Giulietti, AM; Vírginio, J; Gamarra-Rojas, CFL, editors. *Vegetação e Flora da Caatinga*, Associação Plantas do Nordeste. Centro Nordestinos de Informação sobre plantas. Recife, PE; 2002; pp. 49–90 (In Portuguese).

Schmid, B; Balvanera, LP; Cardinale, BJ; Godbold, J; Pfisterer, AB; Raffaelli, D. et al. Consequences of species loss for ecosystem functioning: Meta-analyses of data from biodiversity experiments. In: Naeem, S; Bunker, D. E; Hector, A; Loreau, M; Perrings, C; editors. *Biodiversity, ecosystemfunctioning, and human wellbeing: An ecological and economic perspective,* Oxford: Oxford University Press; 2009; pp. 14–29.

Schleifer, K.H. Microbial diversity: facts, problems and prospects. *Systematic and Applied Microbiology*, 2004, 27, 3–9.

Schnitzler, A; Hale, BW; Alsum, EM. Examining native and exotic species diversity in European riparian forest. *Biological Conservation*, 2007, 138, 146–156.

Schnitzer, S; Klironomos, J; Hillerislambers, J; Kinkel, L; Reich, P; Xiao, K; Rillig, M; Sikes, B; Callaway, R; Mangan, S; Van Nes, E; Scheffer, M. Soil microbes drive the classic plant diversity–productivity pattern. *Ecology*, 2011, 92, 296–303.

Schreiner, R. P., K. L. Milhara, H. McDaniel, and G. J. Bethlenfalvay. 1997. Mycorrhizal fungi influence plant and soil functions and interactions. *Plant Soil*, 188,199–209.

Sekercioglu, CH. Ecosystem functions and services. In: NS. Sodhi, PR. Ehrlich, editors. *Conservation Biology for All*; 2010; pp. 88–104.

Singh, BK; Munro, S; Potts, JM; Millard, P. Influence of grass species and soil type on rhizosphere microbial community structure in grassland soils. *Applied Soil Ecology*, 2007, 36, 147–155.

Siqueira, J.O.; Colozzi-Filho, A.; Oliveira, E. Ocorrência de micorrizas vesicular-arbuscular em agro e ecossistemas do Estado de Minas Gerais. *Pesquisa Agropecuária Brasileira*, 1989, 24, 1499–1506 (In Portuguese).

Smalla, K; Wieland, G; Buchner, A; Zock, A; Parzy, J; Kaiser, S; Roskot, N; Heuer, H; Berg, G. Bulk and rhizosphere soil bacteria communities studied by denaturing gradient gel electrophoresis: plant-depent enrichment and seasonal shift revealed. *Applied and Environmental Microbiology,* 2001, 67, 4742–4751.

Smith, SE; Read, DJ. *Mycorrhizal Symbiosis*. Elsevier, New York; 2008.

Smith M.R., I. Charvat, and R.L Jacobson. 1998. Arbuscular mycorrhizae promote establishment of prairie species in a tallgrass prairie. *Canadian Journal of Botany* 76:947–1954.

Steinaker, DF; Wilson S.C. 2008. Scale and density-dependent relationships among roots, mycorrhizal fungi and collembola in grassland and forest. *Oikos* 117: 703–710.

Stevens, KJ; Wellner, MR; Acevedo, MF. Dark septate endophyte and arbuscular mycorrhizal status of vegetation colonizing a bottomland hardwood forest after a 100 year flood. *Aquatic Botany*, 2010, 92, 105–111.

Strasser, E; Lugo, MA; Perino, E; Fernández, E. Estudio con fines geoquímicos del suelo rizósferico de Poaceae en pastizales de la Puna jujeña. I. III Congreso Argentino de Química Analítica. Merlo, San Luis; 2005; pp. 8–15 (In Spanish).

Suttle, KB; Thomsen, M; Power, M. E. Species interactions reverse grassland responses to changing climate. *Science*, 2007, 315,640–642.

Systematics Agenda 2000 (SA2K). *Systematics agenda 2000: charting the biosphere*. American Society of Plant Taxonomy/Society of Systematic Biologist/W. Henning Society/Associations of Systematics Collections, New York; 1994.

Tabarelli, M; Silva, JMC. Áreas e ações prioritárias para a conservação da biodiversidade da Caatinga In: Leal, I.R., Tabarelli, M. and Silva, JMC, editors. *Ecologia e Conservação da Caatinga*. Recife: Ed. Universitária da UFPE; 2003; pp. 777–796 (in Portuguese).

Tiessen, H; Salcedo, I.H; Sampaio, EVSB. Nutrient and soil organic matter dynamics under shifting cultivation in semi-arid North-Eastern Brazil. *Agric. Ecosyst. Environ*. 1992, 39,139–151.

Torpy, F.R;Morrison, DA;Bloomfield, B.J. The influence of fire frequency on arbuscular mycorrhizal colonizationin the shrub *Dillwynia retorta* (Wendland) Druce (Fabaceae). *Mycorrhiza,* 1999, 8, 289–296.

Torsvik, V; Goksøyr, J; Daae, F.L. High Diversity in DNA of Soil Bacteria. *Applied and Environmental Microbiology,* 1990, 56, 782–787.

Urcelay, C. Co-occurence of three fungal root symbionts in *Gaultheria poeppiggi* DC in Central Argentina. *Mycorrhiza,* 2002, 12, 89–92.

Urcelay, C; Tecco, PA; Chiarini, F. Micorrizas arbusculares del tipo Arum y Paris y endófitos radicales septados oscuros en *Miconia ioneura* y *Tibouchina paratropica* (Melastomataceae). *Bol. Soc. Argent. Bot.*, 2005, 40, 151–155 (In Spanish).

Velazquez, MS; Cabello, MN, Irrazabal, G; Godeas, A. Acaulosporaceae from El Palmar National Park, Entre Ríos, Argentina. *Mycotaxon*, 2008, 103, 171-187.

Velázquez, MS; Biganzoli, F; Cabello, MN. Arbuscular mycorrhizal fungi in El Palmar National Park (Entre Rios Province, Argentina) – a protected reserve. *Sydowia*, 2010, 62, 149–163.

Walker, C; Mize, W; Mc Nabb, HS. Populations of endogonaceous fungi at two populations in central Iowa. *Can J. Bot.*, 1982, 60, 2518–2529.

Wang, B; Qiu, YL. Phylogenetic distribution and evolution of mycorrhizas in land plants. *Mycorrhiza*, 2006, 16, 299–363.

Wang, M; Chen, JK; Li, B. Characterization of bacterial community structure and diversity in rhizosphere soils of three plants in rapidly changing salt marshes using 16S rDNA. *Pedosphere*, 2007, 17, 545–556.

Weber, OB; Souza, CCM; Gondin, DMF; Oliveira, FNS; Crisóstomo, LA; Caproni, AL; Saggin Júnior, O. Inoculation of arbuscular mycorrhizal fungi and phosphatefertilization on dwarf cashew seedlings.*Pesq. Agropec. Bras*., 2004, 39, 5, 477–483.

White, JA; Tallaksen, J; Charvat, I. The effects of arbuscular mycorrhizal fungal inoculation at a roadside prairie restoration site. *Mycologia*, 2008, 100, 6-11.

Whitman, WB; Coleman, DC; Wiebe, WJ. Prokaryotes: the unseen majority. *Proceedings of the National Academy of Sciences of the United States of America,* 1998, 95, 6578–6583.

Zangaro, W; Nisizaki, SMA; Domingos, JCB; Nakano, EM. Mycorrhizal response and sucessional status in 80 woody species from south Brazil. *J. Trop. Ecol*., 2003, 19, 315–324.

In: Native Species
Editors: L. Marin and D. Kovač
ISBN: 978-1-61470-613-7

Chapter 2

The Rise, Fall and Restoration of a Native Breed of Domestic Cattle: The Devon Red Ruby of Southwest England

R. Trevor Wilson
Bartridge Partners, Bartridge House,
Umberleigh, Devon, UK

Abstract

Early British explorers and settlers introduced their domestic livestock to new lands. Many species and breeds were indigenous to the place of origin of the introducers. The Devon breed of cattle is one example of such introductions. Cattle recognizable as the progenitors of the Devon type appeared in its native county in southwest England in Domesday times at the end of the eleventh century and a beast of Devon colour was given as heriot to Tavistock Abbey in the mid-fourteenth century (Finberg, 1951). There is no direct evidence of a Devon "breed" before the eighteenth century but a type was beginning to be selected and fixed by local stockmen from this period on. A Herd Book (pedigree records) was established in 1851 and a Breed Society formed in 1884. Early in the twentieth century the Devon was the second most numerous breed in the UK. Numbers declined after the Second World War in particular with the influx of "continental" breeds that were introduced in response to changing customer demand for meat but have since recovered to some extent. Red cattle from Devon were taken to the USA by pilgrims early in the seventeenth century and to Australia some two hundred years later. Devons, as such, have been exported to about 30 countries with numbers varying from as few as one to more than 800. In some countries they were used for crossbreeding and especially to produce beef from dairy herds. In others, notably Australia, Brazil, New Zealand, the USA and until recently South Africa they have been maintained as pure breeding herds with their own Herd Books and Breed Societies. In the early twentyfirst century the Devon, bred for thriftiness, fertility, docility and easy management is ideal as a beast that provides quality meat to a hungry world concerned about economic and environmental costs.

Keywords: Indigenous Domestic Animals; Breed Development; Eminent Stockmen; Livestock Introductions; Exports; Crossbreeding

INTRODUCTION

British exploration of new lands and its natural corollary of settlement started to turn the world map red as early as the sixteenth century. On many early voyages livestock were taken to provide fresh meat for the crews and domestic animals – poultry, pigs and to a lesser extent sheep and goats – were often released at landing points and left to multiply for harvesting on later visits. At a later stage British colonists and settlers took livestock on their migrations and established breeding herds in all the countries and on all the islands where they set up home. Many introductions were local breeds of all species. Even as late as the middle of the twentieth century livestock of British origin retained their pre-eminence and only three of the top twenty cattle, sheep, pig and horse regarded worldwide as having most merit were not of British origin: these non-British animals were Friesian cattle, Merino sheep and the Percheron horse (Scott Watson and Hobbs, 1951).

The County of Devon in southwest England has always been a pastoral as opposed to an agricultural area. In 2006 permanent grassland occupied 50 per cent of farms in the region compared to 36 per cent in the country as a whole. The South West Region, reflecting the importance of grassland, also had a higher proportion of lowland livestock farms (21 per cent) than the national average (17 per cent) (DEFRA, 2006). Within the generic "grassland", however, there is a wide range of habitats ranging from sea level marshes and meadows through mid level short and medium term leys to upland moors of fine grasses and heather. Within this rich ecological region there is a concomitant richness of domestic livestock species and types. Devon is a major repository of farm animal diversity and is home to at least five breeds of sheep (Devon Closewool, Devon (and Cornwall) Longwool, Exmoor Horn, Greyface Dartmoor, Whitefaced Dartmoor), two of ponies (Dartmoor, Exmoor) and two of cattle (Devon, South Devon). All the sheep and both pony breeds are at some degree of risk of extinction on numerical or restricted geographical area grounds (RBST, 2009; Wilson, 2011)

This paper chronicles the origins and history of the Devon breed of cattle – also sometimes known as Beef Devon, North Devon, Red Devon and Red Ruby (the Devon should not be confused with the South Devon and, although one authority (Mason, 1996) states the alternative names are not recommended for use in English the British Devon Cattle Breeders' Society promotes "Ruby Red Devon" as a brand name) – and its global spread and influence.

A Native and Adapted Breed: Origins and History

> "The origin of most communities is necessarily obscure; it is, therefore, not to be wondered at that the fragments of history on which the origin of a particular breed of cattle rests are somewhat shadowy and uncertain." (Tanner Davy, 1869).

Devon has always been a pastoral county. As early as the Domesday period, northwest Devon in particular had a dense population of cattle which, at 7.2 head per 1000 acres (405 ha), was twice that of the county as a whole (Morgan, 1940). The first indication of a red animal clearly a prototype Devon appears in the mid fourteenth century – when references to the colour of domestic stock were few – when a red bull came into Tavistock Abbey in 1366 as a heriot from a tenant. Before the end of the seventeenth century the progenitors of the Devon in the area of Great Torrington were considered to be "as good as any in the Kingdom" (Risden, 1957). There is much evidence for a concentration of cattle in northwest Devon in the area of Torrington-Bideford-Barnstaple-South Molton (it should be noted that in the twentyfirst century as well as in Devon there are eponymous cattle at Barnstaple (Barnstable) in the USA and in Australia the registered name of one herd is "Barnstaple" in honour of the English town and because the owner is called Barnes) spanning the period Domesday to the end of the seventeenth century. This gives an unbroken ancestry of 600 years and provides a lineage as ancient as any

British bovine. These cattle were still, however, small at the end of this period as naval purveyors needed to fix the minimum weight of a fat ox – of probably 4 to 5 years of age – at 6 cwt (672 lbs, 305 kg) (Hamilton, 1878; Trow-Smith, 1957).

The Devon, as such, was well known to fashionable and progressive breeders towards the end of the eighteenth century. The elite herd of the time was that of the Quartly family of Great Champson farm at Molland on the southerly slopes of Exmoor. This herd was established by James Quartly in 1776 although the family had farmed here for one hundred years before that and took the lease of the farm in 1703. On his death in 1793 James left the lease of the farm and the herd to his third son Francis who began systematic improvement of the breed in 1794. The two older brothers also bred Devons and were considered good stockmen and judges of animals. During the wars with France and the USA at the beginning of the nineteenth century many of the best animals were sold for beef to feed the armies, resulting, in many cases, in degeneration of the stock.

Portrait by Thomas Mogford, reproduced by courtesy of Mrs M Turner.

Figure 1. Francis Quartly with Cherry 66 and calf.

Francis Quartly set about improving this situation by buying in some of these animals, outbidding the butchers, and -- as one noted Devon breeder remarked -- taking the trump in every hand and retaining them for breeding (Tanner Davy, 1869). Many Quartly females were thus bought in but Francis used his own bulls in the herd. Already at this early stage he could command five guineas as a service fee for covering other breeder's cows (Hall and Clutton-Brook, 1989: Thornton, 1992). Quartly bred for meat and draught with milk production being only a minor consideration. When Davy's Devon Herd Book was first published Quartly was acclaimed as the saviour of the race and the founder of the true breed. Just prior to the Herd Book's publication a portrait of himself and two of his animals was presented to Francis Quartly in recognition of his achievements (Figure 1) (Beer and Beer, 1998).

It was probably the dual purpose nature – draught and meat – of eighteenth and nineteenth century Devons that made them popular with the nobility and the gentry: far from being "old fashioned", the use of oxen for work was still an important asset. The Earl Fortescue and Baron Clinton, both Devon land owners, had herds. His Royal Highness the Prince of Wales (who is also the Duke of Cornwall and owns vast tracts of land in Cornwall, the county next to Devon) had a herd on the Duchy Home Farm. A plethora of other gentry who occupied several pages in Burke's Peerage also rode on the band wagon. Farther afield the Duke of Bedford owned Devons as well as the noted agricultural innovator Thomas Coke, Earl of Leicester who kept a herd at Holkham Hall in Norfolk. There was also a number of herds in Ireland. In addition to the admitted usefulness of the Devon a further possible reason for the landed gentry keeping them at this stage was that, along with the Sussex and to a lesser extent the Hereford, they were the breed in which the continental influence was then least apparent (Hall and Clutton-Brook, 1989) and thus they could flaunt their "patriotism".

If it were the Quartly family and other progressive local husbandmen that were the master craftsmen in fashioning the designer breed that we recognize today, it was another Devon farming family that put order into it. John Davy, who was born in 1706, died in 1790 and left his Devon herd to his sons John Tanner Davy at Rose Ash about 10 miles from Molland and William Davy at Flitton Barton about five miles to the west. John Tanner Davy died in 1852 and was succeeded by his son of the same name. William promoted the dairy properties of his cattle with considerable success but never lost sight of the distinctive beef characteristics of the breed. The Flitton Devons won many top show prizes and had a considerable influence on the breed as a whole. It was the second John Tanner Davy (Figure 2), however, to whom the Devon breed owes most as it was he who published the first Herd Book (Tanner Davy, 1869: Thornton, 1993).

Volume 1 of Davy's Devon Herd Book was published in 1851 in time for the Royal Agricultural Show which was held that year at Windsor. This first volume contained the pedigrees of 132 bulls and 483 cows arranged alphabetically and numbered separately for the sake of distinction and reference. Davy's Devon Herd Book Volume 4 was published in 1863 which, inclusive of the earlier editions, brought details of 768 bulls and 2474 females on to the record (Tanner Davy, 1869)[1].

From Volume 1 of the Herd Book, 1880 reprint; courtesy of Devon Cattle Breeders' Society.

Figure 2. John Tanner Davy, founder of Davy's Devon Herd Book.

[1] Tanner Davy's account of Devon cattle history in the Journal of the Royal Agricultural Society of England is essentially a narrative of pedigrees, ownership, shows and prize winners and sales and prices of many of the Devon cattle herds of the mid nineteenth century.

The breed gathered momentum from this period, began to spread more widely throughout the United Kingdom and provided serious competition for the ubiquitous Shorthorn. Herd Books appeared intermittently until the Devon Cattle Breeders' Society (DCBS) was formed in 1884. An early action of the DCBS was to acquire the rights to the Herd Book from Messrs Hawkes and Risdon, Auctioneers of Williton to whom Davy had sold the copyright, and to purchase all stocks of earlier editions, for which they paid the princely sum of £20 (Thornton, 1993). Henceforth the Herd Book was published annually. The next agenda item for the duly constituted committee was that imperative of all pedigree breeding, the establishment – separately for males and females – of a Standard of Excellence. This was followed by the next priority, the organization of the Society's own shows and sales. The Standard of Excellence has apparently stood the test of time and, one and a quarter centuries after it was first promulgated, remains little changed from the original (DCBS, 2009) although beauty remains firmly in the eye of the beholder as a series of visits to the contemporary show ring by an objective observer will demonstrate.

The next 25 years, spanning the changeover from the nineteenth to the twentieth centuries, was a golden age for the Devon. It spread widely throughout the United Kingdom (and overseas, see next section) from its southwestern homeland. By 1908, as shown by a national census of agriculture and to the surprise of many, it was the second most numerous breed in the United Kingdom. At 454,694 head it had, admittedly, only about 10 per cent of the numbers of Shorthorns (4,413,040 head) but was ahead of the Hereford (384,877) by 70,000 head and the Aberdeen Angus (193,960) by more than a quarter of a million animals (BOAF, 1912). These numbers included all animals of a breed type but in terms of registered pedigree animals the Devon was seventh in a league table of nine with 5.2 per cent of the 115,618 registered pedigree cattle in England and Wales being Devons (MOAF, 1921).

The Devon remained popular throughout the first half of the twentieth century. It maintained its place numerically as well as its eminence in the major show rings in interbreed competitions and at fat stock shows at Smithfield and elsewhere. An attempt was made to promote the Devon as a dairy animal during early years of the twentieth century. Classes for Milking Devons were held at various major agricultural shows and Devons were entered in interbreed milk classes at Smithfield during the decade 1911-1920 and a special section was introduced in the Herd Book. Long noted for milk quality – butterfat and solids-not-fat (think Devon Cream Teas!) – rather than quantity, there was little support for this initiative and any impetus soon faltered. Decline in breed numbers and membership set in during the 1950s and 1960s. The increased demand for milk and dairy products, changing consumer preferences from fatter to leaner meat, demand from the emerging supermarket chains for larger joints and government's decision to allow imports of continental breeds were principal contributors to this decline. By the late 1970s and early 1980s the breed and its Society were encountering serious difficulties. The number of members was very low and registrations fell dramatically.

Rearguard actions were taken by a few dedicated breeders and DCBS members to redress the situation. In 1954, after years of heated debate and over the objections of die-hard traditionalists, the decision was taken to introduce a polled strain. Polled Devons originated from a naturally hornless bull born in the USA in 1925. Unfortunately this animal lacked many of the breed's perceived qualities but over the years a naturally polled strain was developed. American stock was imported to use on British Devons but bulls from other polled breeds, notably Red Angus, Red Galloway and Red Poll, were also used (Thornton, 1993). To accommodate this momentous step a polled section was admitted as a separate entity in the Herd Book. Other categories were also introduced into the Herd Book in order to broaden the genetic base of the breed and to "improve" it. Thus, over time, there have been a Supplemental Register, Extra Supplementary Register and Experimental Grading-Up Register in addition to the Polled Register: of these the first two still exist in 2011 whereas polled animals have been fully integrated into the main register where they are distinguished by a letter 'P' as an alphabetical prefix to their numerical identity. These ploys allowed, initially, purebred but non-pedigree females and females considered of Devon type on inspection eventually to gain admission to the main register. They also allowed the blood from other breeds to be introduced in attempts to put size and better form into the Devon. The most favoured of these other breeds was the Salers from France which is an animal very similar in type and colour to the Devon and with which there had already been some to and fro crossing during the nineteenth century (Sinclair, 1893). Other French breeds having nothing in common with the Devon, notably the Limousin and Charolais, were also used in attempts to increase the scale of the native breed as can be ascertained from several of the Society's Herdbooks.

The "outcrossers" were, however, in the minority of Devon breeders. Most preferred to attempt breed improvement with a view to meeting market demands by in-breed selection. Most of these, again, did not stray from the native breed in its native country but some resorted to importing semen from

Devon cattle established overseas and, notably, from New Zealand[2]. In essence, the purebred Devon has remained "pure". It is to the credit of the breeders and the DCBS that in holding on to this purity the twentyfirst century Devon has evolved as a beast capable of fulfilling modern market demand whilst continuing to maintain its ability to produce a quality product at low financial cost mainly on grazing alone and at least cost to the natural environment.

A Global Player: Geographical Spread and Influence

Almost 400 years have passed since, one third of the time through the seventeenth century, there is proof that protoDevon cattle were exported from their own tract. The Plymouth Plantation or Colony was established in New England in 1620 but did not import "neat" cattle until 1624. In that year "Mr Winslow came over and ...... brought 3 heifers and a bull, the first beginning of any cattle of that kind in the land ....". In 1625 a further shipment of cattle arrived in America together with cloth and other goods with "the cattle [being] ye best goods, for ye other being ventured ware, were neither at ye best (some of them) nor at ye best prises." These cattle, which were originally public property under the idealistic constitution, were later divided among the settlers who had become disenchanted with the system, at the rate of six persons (or shares) per cow (Deane, 1851). Four years later Edward Winslow "sold unto Capt. Myles Standish his sixth share in the red cow in consideration of 5 pounds 10 shillings". Multiply that by 400 years of inflation to arrive at a modern value for that "red cow"!.

There were few imports from Europe after 1630 and the Plymouth colony bought stock from Virginia. Even at this early stage of development the hardiness and thriftiness of English stock was recognized as, although smaller and less milky than the Dutch animals, they were "much less trouble than

---

2 In this respect it has adopted an approach different to that of the Hereford and Aberdeen Angus breeders who resorted to massive imports of American and Canadian blood – in which considerable outcrossing had been used – to increase the size of their animals. This modified the breeds to such an extent that the British Rare Breeds Survival Trust felt obliged to encourage separate Herd Books for "original populations" of the Aberdeen Angus which has less than 150 registered adult breeding females and is placed in the Critical category for conservation and the Hereford with less than 1500 registered adult breeding females and is in the Minority category (RBST, 2009).

those of Holland, as most of the time, if any care be requisite, it is only for the purpose of giving them occasionally a little hay" (Jameson, 1909).

At least one authority on the history of early cattle imports to New England considered that those introduced by the Pilgrims were indeed of Devon type. Thus:

> "It is logical to assume that cattle shipped from Plymouth and Barnstable [sic] came from the surrounding Devonshire area. In this area the Devon breed of cattle was developed. As there is no record of any mass movement of cattle to or from Devonshire during or following the colonization of North America it may be concluded that the first cattle brought to the New England colonies were of an inheritance similar to the Devon breed that was later developed and improved in Devonshire. The fact that Devon cattle have always been rather popular in the New England States adds further to the belief that the first cattle imported were of the Devon type." (Bowling, 1942).

There is no doubt that cattle from Devon were taken abroad later in the seventeenth and throughout the eighteenth centuries. It was to be another two hundred years, however, before there is further confirmation of Devons exported from England. William Duckitt was sent out to South Africa by the British Government in 1800 with a view to improving agriculture there. His diary notes that he considered oxen there to be weak because of their long legs and to improve that trait he had "taken out a Devon bull and two heifers from the Duke of Bedford's prize stock." There do not appear to be any further records of these animals (SASB, 1971).

Information on early exports of Devons to Australia appears in the Van Dieman's Land [Tasmania] Annual for 1834. It says the first cattle arrived in 1807 and were sold to settlers by government on long term credit. It was not, however, until 1825 or 1826 that there were direct imports to Van Dieman's Land "of superior bred animals from England [that were] an object of attention with certain emigrants who were embarking hither, so that, by degrees, the old original breed has now become almost lost. We have had pure Devons, Herefords, Durhams, Holdernesses, Fifeshires, and others of the most admired breeds among the English farmers, brought here in considerable numbers". Further, "Large importations of their best and improved crosses have for some years previously, from time to time, been made from New South Wales; so that many individuals had been possessed of herds, very different from the original stock, even before these English importations reached us. At present, either for the yoke or the pail, for docility or for hardiness, the improved breed of cattle, which is rapidly taking place of all other, cannot be surpassed, either in England or in any other part of the World." James Bryant arrived in the colony on 4 February 1825 and was joined in October by his family accompanied by a 2-year old Devon bull and a heifer of the same age. The Hobart Town Gazette of 31 December 1825 advertised "for the convenience of the Gentlemen of Hobart Town, and the neighbourhood; for two months only, a beautiful Devon bull, two years old, at two guineas each cow - application to be made to Mr. Edmund Bryant, Rock House, Campbell Street; or to Mr. James Bryant, at Jerico". This bull was one of eight Devon cattle, described as thoroughbred, that arrived at Hobart on 24 October 1825. Somewhat later, on 7 May 1826, George Hobler arrived in Hobart with his wife, servants and a Devonshire heifer known as Fairmaid. In his application for a land grant Hobler states that he had "secured some cattle of the highest blood in the North Devon breed so valuable for draught" (DCBSA, 2009).

The first recorded import of Devons to New Zealand was in 1836 by James Busby, British resident for the Queen, stationed at Waitanga in the Bay of Islands. A group of 20 heifers and a bull were sent from England and the resulting progeny formed the basis of the bullock teams that were vital to land development. Drovers loved them for their docility and their sweeping horns which helped define the width on the primitive tracks. Devons were also used to break in and cultivate land on the first Mission Station established at Waimate North also in the Bay of Islands. In addition to draught power these early animals also provided milk and meat for the early settlers (NZRDCBA, 2009).

By the mid-1860s, according to Davy, Devons had been exported to France, Denmark, Mexico, Jamaica, Canada, the United States of America and Australia (Tanner Davy, 1869: Sinclair, 1893). A notable exception from this list is New Zealand whither we know Devons went as early as 1836 (see preceding paragraph). South Africa is missing because the first exports there, after the very early ones of 1800, were in 1872. For one consignment to Japan, in addition to the usual selection criteria, the buyers insisted on firm bone and sound feet (DCBS, 1911). Over succeeding years Devons were "officially" exported to upwards of 30 countries around the world (Figure 3)[3]. Brazil has been the main destination of UK exports, followed by South Africa (including, prior to its constitution as the Union of South Africa in 1910, to at least two of its constituent colonies, Cape of Good Hope and Orange River) and then Australia, the USA and Rhodesia. Notable also is the very small number of

[3] Data for Figure 3 (and for Figure 4) are extracted from DCBS Herd Books for 1901 to 1990 under the rubric Export Certificates Granted.

animals exported directly to New Zealand which now has a high profile in the Devon world. Several countries that imported one or very few animals probably did so for two very different reasons: the one for use in experimental work on crossbreeding or to improve a specific trait in their national stock and the other by an enthusiast for the breed.

There were also unofficial or private exports that apparently did not pass through the DCBS. There was a Devon bull in Tanganyika Territory [Tanzania) from 1926 to 1928 which was presumably to be used on the local Iringa Red zebu but there is no information whence it came nor indeed whither it went (DVSAH, 1926: 1927: 1928). Another example is Russia. The Tambov Red breed was created in the middle of the nineteenth century by crossing the local improved cattle in Tambov province with the Tyrol (also known as the Zillertal) breed with some additions of Devon and Simmental blood. The Tambov was not actually recognized until 1948 when a Herd Book was established. In 1980 the total population was 45,000 head. Its characteristics are seen as adaptation to the ecological and economic conditions of Tambov Region, a strong constitution and resistance to economic diseases.

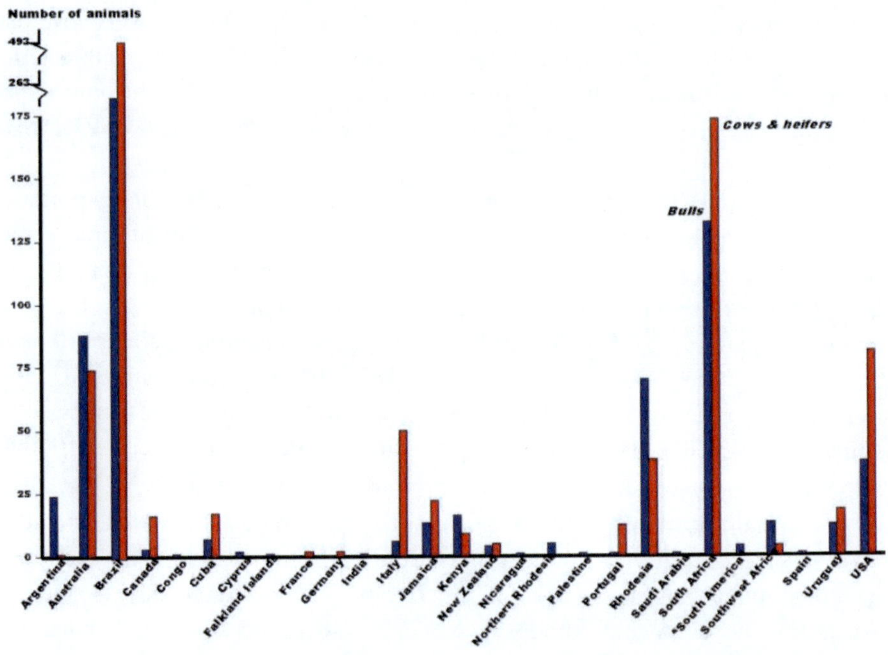

Figure 3. Numbers of Devon cattle exported to various countries from UK, 1901-1990.

Under standard conditions of breeding and management the Tambov has higher milk, fat and solids-not-fat yields than the Simmental which under the former Soviet system was the planned breed for the region: fattened bullocks weigh 1100-1250 kg (Dmitriev and Ernst, 1989: Mason, 1996).

Between 1901 and 1990 some 1751 Devons were exported through "official" channels, comprising 722 bulls and 1029 cows and heifers (Figure 4). Most animals were exported just prior to and just after the First World War when bulls seemed in greatest demand and then again in the 20-year period following the Second World War when it was females that were mainly sought. Brazil, in addition to being the country that provided greatest demand, imported stock over a very long period. Most demand from South Africa was in the peri-First World War period with smaller numbers then imported regularly until imports ceased altogether in 1971. Demand from Australia continued for most of the first half of the twentieth century until direct imports of live animals were prohibited altogether in 1958 (although it appears derogations were granted for further imports in 1980, 1981 and 1985). The USA was also a regular customer for home country Devons until the end of the 1970s[4].

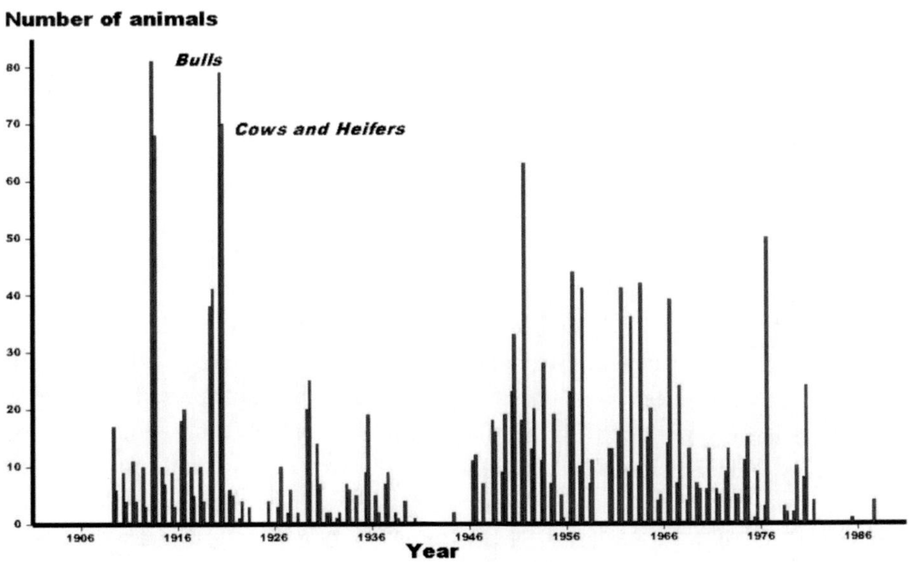

Figure 4. Numbers of Devon cattle exported from UK, 1901-1990.

[4] Data were compiled for this paragraph from various volumes of the DCBS Herd Book.

UK exports of live Devons slowed in the 1970s and came to an effective halt in the 1980s. A principal reason for this was the increasing use of artificial insemination with semen being first exported to Brazil in 1968. A second major reason was the risks – perceived or real – associated with live animal movements across international boundaries[5]. By this time, nonetheless, Devons were well established as pure breeds in several countries. In the USA a Herd Book was initiated in 1855 and a Breed Society formed in 1905, Australia started a Herd Book in 1873 and a Breed Society in 1929, South Africa's Breed Society was formed in 1917 (Devons had been recorded in the South African Stud Book started in 1906 prior to the formation of the Society), Brazil started a Herd Book in 1952 and New Zealand where there was a very late start to their Herd Book in 1964 and an Association was not formed until 1972 (Mason, 1996).

The beginning of American Devons was the three heifers and one bull taken to Plymouth Plantation by Devonshire pilgrims. Devons, plural, because it is only in the USA that two separate types of (Red Ruby) Devon are maintained. Their immediate value, on first import, was as draught animals but the little milk the cows produced surplus to that taken by the calf was here, as in the beasts' home tract, valued for its cream and for the butter and cheese made from it. At the end of its working life the ox produced an adequate beef carcass. Just as in the UK, where Queen Victoria and her direct descendants bred Devons, so in the USA the top people were enamoured of Devons: George Washington was an early Devon breeder, the descendants of whose herd can still be seen at his home at Mount Vernon and Henry Ford who was a later convert. Since those very early days there has been a steady trickle of pure Devons from England to the USA with 37 bulls and 81 females imported between 1929 and 1978. There has also been reverse flow, most notably of polled animals which arose from a naturally polled male born in the USA.

The Milking Devon became well established in New England during the late seventeenth century and spread down the coast as far as Florida during the eighteenth and nineteenth centuries. By the end of the latter period, however, it had been largely replaced by the Shorthorn and by 1900 the Milking Devon was rarely seen outside New England. As the market for dual-purpose cattle dwindled the Milking Devon virtually disappeared. By the mid 1900s, and in response to this loss, Milking Devon breeders began selecting for beef. The

---

[5] Bovine Spongiform Encephalopathy (BSE, "mad cow disease") was widespread in the United Kingdom in the 1980s but was mainly confined to intensive dairy rather than to more extensive beef operations.

new population became known as the Beef Devon. A few select traditional breeders nonetheless continued multipurpose selection and formed the American Milking Devon Association in 1978. This resulted in what is now the American Milking Devon which is claimed by its owners as the closest relation of the original Milking Devon. Small in numbers – in the year 2000 there were totals of 119 bulls and 492 cows -- this Devon type is listed as critically endangered by the American Livestock Breeds Conservancy (Splan and Sponenberg, 2004: AMDA, 2009).

In "modern" times Devons were imported to Plymouth, Massachusetts USA by Messrs Winthrop in about 1800 and in 1805 to Otsego County, New York by General Eaton but nothing further is known of these animals (Sinclair, 1893). In 1817 Coke of Holkham (later Earl Leicester) presented six heifers and a yearling bull named Taurus from his celebrated herd in Norfolk (Figure 5) to Robert Patterson of Sykesville, Maryland and in 1834 Patterson's brother obtained another bull, Anchises, from Coke (Sinclair, 1893)[6]. By 1844 some 13.4 per cent of the 186 Devons in the USA were direct imports and pedigree Devons represented 9.1 per cent (Shorthorns 89.6 per cent) of all pedigree cattle in the USA. Devons maintained a position in the pedigree stakes until 1870, by which time only 3.4 per cent of the national total of 3292 pedigree cattle were Devons with this breed then occupying third place in the US herd after Shorthorns (77.0 per cent) and (Jerseys 9.9 per cent) but still well ahead of the Hereford (0.5 per cent) as the only other beef breed. In 1921, however, Devons had been relegated to "other beef" breeds (6.1 per cent) after Shorthorn (39.2 per cent), Hereford (38.1 per cent), Aberdeen Angus (10.2 per cent), Polled Durham (5.8 per cent) and Galloway (0.7 per cent) when beef breeds represented 53.7 per cent of all pedigree cattle and dairy breeds 46.3 per cent (US Census, 1920: Brinkman, 1964: Walton, 1999). The first Devon Herd Book in the USA was begun in 1855 – prior to this American Devons were registered in England – by Horace Sessions in Massachusetts: the last volume of this series appeared in 1879. Two years later James Buckingham of Zanesville Ohio began publishing the American Devon Record and this registry has remained intact for 130 years. Unlike the case of the UK and most other countries (but see the paragraph on New Zealand) the American Devon Cattle Association does not allow upgraded animals in the main register. The Percentage Registry programme provides that cattle may be registered and bred up to 15/16 Devon as females or 31/32 as bulls before they gain full

[6] Sinclair quotes a letter from G. Patterson to R. Peters of Atlanta, a copy of which appeared in Volume 2 of Davy's Herd Book..

blood status. Even then, however the registries are not co-mingled but kept separate to preserve the breed's genetic integrity. To ensure that animals in the Percentage Registry are distinguishable from pureblood animals, registrations are printed on different paper and have different names and seals. Animals in the Qualified Registry will never be in the same registry as the pureblood animals which must have a pedigree verifiable back to England where the breed originated. Devon bulls are important in crossbreeding in the beef industry's move from the meat of the 1980s and 1990s back to the type of beef prevalent in the 1940s and 1950s. Because Devon breeders were immune to the growth and feedlot culture of the post WWII era the breed has retained its compactness, foraging ability and carcass merit. Devons today have "once again become the all-American cow" because of their ability to perform and finish on grass, superior fertility and maternal traits and most of all their ability to deliver high quality beef to the consumers. In the USA Devons are successfully used in the temperature extremes of Nevada, Montana, Wyoming, New Mexico and Maine, in the swampy and unhealthy areas of the Gulf States and in the rocky woodland hills of cold northern New England (ADCBA, 2009).

Figure 5. Thomas Coke of Holkham, Norfolk and a prize ox of the type exported to the USA in the early nineteenth century.

Some years after the first imports of Devons to Tasmania the Trethewie family arrived there from Truro in Cornwall England and started their herd with stock acquired from Edmund Bryant, George Hobler, a Mr Manley and

W.N. Whittle. George Hobler left Tasmania for Sydney in 1836, taking with him some choice Devons. Charles Reynolds arrived in Australia in 1829 from Raddon Court near Exeter in Devonshire. In 1842 he purchased ten imported cows and then got James Quartly to select two bulls which arrived the following year. These were followed by an impressive series of males and females from England and additional animals from Tasmania. Reynolds bought 300 of Hobler's best cattle in 1843. The Reynolds herd was to become the base for breed expansion on mainland Australia. Further Devons were imported to Australia in 1850 by the Codrington brothers of Wiltshire, England. The Whittle family had originated from Dorset and had imported a bull who appeared in the first volume of Davy's Devon Herd Book. A John Lord founded his herd in 1851 with a bull and a cow also imported from England. The Trethewie family also imported many Devons direct from England and eventually sent many cattle to New South Wales and on 4 May 1875 shipped 40 cows and heifers, one yearling bull and four calves there. In the last quarter of the nineteenth century there were many more imports from England and many new and long-lived herds were started: one existing property has had Devons roaming it for more than 120 years. Between 1909 and 1958 (when live animal imports were stopped) 94 bulls and 74 cows/heifers were imported thus providing adequate new blood for the animals already present. Devons were still to the fore in the country in 1917 and did well at the various agricultural shows. At a large ranch in Queensland at the beginning of a long drought about this time one owner had 12 Devon bulls and 60 of another breed whereas afterwards he had ten Devons but only two of the other breed (DCBS, 1917: 1921). In the 1930s Devon bulls were used on halfblood Zebu-Hereford crosses in an attempt to produce a quarterblood Zebu that retained tick resistance and hardiness in the northern Australian environment but produced a better carcass more quickly than the Zebu or halfblood. Towards the end of the 1940s Devons were crossed with the locally predominant Shorthorns to produce breeding bulls for the harsh conditions and remote areas of northern Queensland (Kelley, 1959). The Australian Sahiwal mainly centred in Queensland was developed from ten imported Sahiwals that were used to upgrade Shorthorn, Devon and Jersey cattle to at least 7/8 Sahiwal (Mason, 1996). In the early twentyfirst century the Devon Cattle Breeders Society of Australia has more than 80 members of whom 58 are full and 23 are commercial. The Devon breed has 2285 sires and 11,827 dams in the Group Breedplan database. Several herds are large to very large ranging from 150 to 700 head. Most existing herds are in New South Wales but the Breeders Society hopes that there will soon be Devons over a much wider area. The wide genetic base coupled with a strong phenotype will allow the Devon, with an emphasis on polled animals that are highly fertile, efficient in the use of feed and show good carcass quality, to continue to be a major environmental, economic and social force in the Australian beef industry both as a pure bred and as a terminal sire (Beer, 2009: DCBSA, 2009).

There were no known imports to South Africa after 1800 until 1872. In that year William R. Southey imported one bull and two cows that had been selected for him by his father Sir Richard Southey (one time Colonial Secretary but born at Culmstock in Devon and probably on terminal leave from his official post at the time) from a well known English herd. In later years Southey imported many more Devons (SASB, 1971). The first entries for Devon cattle in the South African Stud Book were in Volume 1 issued in 1906 and comprised two bulls and 18 cows/heifers, all apparently imported, and all the property of W.R. Southey. South Africa imported 132 bulls and 164 cows/heifers between 1909 and 1971 with the Southey family continuing to import stock for many years in to the 1920s. Some consignments were quite large with, for example, 25 bulls and 43 cows/heifers in 1913 (and 46 to Rhodesia), 14 and 19 in 1919 and 131 and 39 (of which 21 and 11 went to the Transvaal Consolidated Land and Exportation Company Limited) in 1920. Devons went to all the states of the Union and "glowing reports" of their success were received (DCBS, various volumes). The Devon Cattle Breeders' Society of South Africa was formed in 1917 and the breed continued to be popular until the 1970s. Although maintained as a pure breed their main role was to supply sires for crossing with the Afrikaner-type females that comprised most of the commercial beef cattle herd and with which they produced good slaughter animals. By the late 1970s structural changes in the beef industry coupled to performance recording (started in 1961) led to re-ranking of breeds. Participation in the scheme was crucial but the Devon failed in this regard with hardly any participation in official performance testing either on-farm or in centralized tests at a time when the Hereford and Angus breeders were rapidly adapting to the new situation. The Devon Breed Society disappeared as animal numbers dropped below the critical level to support it. Only 20-40 Devons were registered per year in the 1980s and the last Devons registered in the South African Stud Book (to which the breed had reverted) were from Namibia [South West Africa] in 1985 and the Eastern Transvaal in 1987[7].

[7] I am indebted for much of the general information on South Africa to Charl Hunlun, Professional Services Manager of the South African Stud Book. The last Devons to be

The Associação Brasileira de Criadores de Devon (ABCD) is the producers' society in Brazil. Devons were first taken to Brazil at the very beginning of the twentieth century. In the succeeding one hundred years cattle were imported almost every year (including two bulls during the Second World War) to a total of 263 bulls and 493 cows and heifers. By 1920, based on the criterion of number of enquiries received, it was hoped by the President of the home society that the Devon would soon be firmly established in Rio Grande do Sul (although in 1917 there had already been a herd of over 1000 pure bred stock there). It is in this subtropical province that Devons have their stronghold but they are also found in the far north. In 2008 the ABCD had almost 300 members (far more than Australia or New Zealand). In recent times approximately 1500 pure Devons per year have been registered together with about 1000 PC (Supplementary Register) animals that are the progeny of registered pure Devon bulls and unregistered but pure Devon mothers. In addition to the normal hazards of climate, high prices for inputs and low prices for outputs, Devon breeders in Rio Grande Do Sul have had to compete against cellulose manufacturers who have bought up prime pastoral land and in the centre and north and against government emphasis on bio-fuel (alcohol) production from sugar cane. In spite of these problems, however, beef producers persevere with, for example, 70 head from 18 exhibitors (including two first time entries) forward at Expointer 2008 (a major agricultural show and to which the Brazilian Society invites judges from all over the globe). During the World Conference held in the USA in 2004 the organizer declared that Brazil was the country with the biggest and probably the best Devons in the world (Cirne-Lima, 2008).

Following the James Busby imports to New Zealand in 1836 there was little immediate activity. Other animals from Australia and from the United Kingdom were imported later in the nineteenth century and the breed continued to grow in popularity until the 1920s when interest waned. Devons continued to survive in small pockets and some established herds maintained breeding lines that traced back to the original Busby cattle. There were then some additional imports but, of all countries in which the Devon has been established as a breed, New Zealand has imported the least numbers and only four bulls and five cows/heifers (with the addition of one batch of semen in 1978) were brought in from England in the period 1911-1981. New Zealand did not establish a Herd Book until 1964. In 1972 a drive to re-establish

registered in 1987 were the property of H.L. Hall and Sons, Mataffin near Nelspruit in the current Mpumalanga Province (old Eastern Transvaal) who had first imported Devons in 1927.

Devons was made and the first Devon Breed Society was formed. It has since been rigorous in maintaining breed "purity". Compulsory performance recording – the first New Zealand breed to do this – was introduced in 1980. From 1990 the national Devon herd was closed to bred-up animals although semen and embryos from approved overseas animals continued to be introduced. The "closed to bred-up" policy created a problem in importing "pure Red Devon" genetics, particularly from England. Here, Salers were used rather widely as an out-cross and the progeny then back-crossed to the Devon and it was predominantly these animals that were presented in England for semen collection for export. In 2008, however, a major change in policy made it possible – albeit with considerable opposition by breeders who were very keen to protect the "pure Devon" status of the New Zealand herd – to register animals with 1/32 non-Devon ancestry. Consequent on this the use of semen from two overseas bulls of 1/32 ancestry was approved along with semen from four "pure" Devons. The New Zealand Red Devon Cattle Breeders Association (Red was added to the original title in order to differentiate the breed from the South Devon) has just over 50 members. About half of these enter calves in the Herd Book from the 600-plus registered cows each year. Devons are reared pure for beef but many bulls are used by dairy farmers for mating heifers or as sweepers after AI with dairy breed bulls. In spite of smallish numbers and what must be the resulting rather narrow genetic base, New Zealand is predominant in exporting Devon genetics to other countries including the USA, Brazil, Uruguay, the Falkland Islands and even back to the United Kingdom. The potential problem of loss of genetic diversity worldwide is exacerbated by the fact that much New Zealand semen used throughout the world is from one not-very-large stud and from very few bulls within that stud: in addition this stud exported 10 in-calf heifers to the USA in 2003 (NZRDCBA, 2009).

Jamaica was amongst the first countries to import Devons in the mid nineteenth century. Imports of small numbers continued throughout the period 1920 to 1963 when a total of 12 bulls and 22 females were introduced. Devons were probably bred pure but their main contribution to Jamaican livestock production was as a foundation breed of the Jamaica Red (also known as the Jamaica Red Poll and the Goodhope Red) together with the Red Poll and zebu *Bos indicus* types. The Jamaica Red is a single purpose beef breed that was apparently sufficiently well typed for a Breed Society to be formed in 1952 but subsequent improvements probably took place with the help of seven Devon bulls and 17 females imported from the UK between 1956 and 1963. In 2008 beef cattle were equivalent to 14.4 per cent of Jamaica's national herd of which 70.6 per cent (or 10.2 per cent of the whole national herd) were Jamaica Red. They thus far outnumber the other beef breeds including the Jamaica Brahman and Jamaica Black but are few in comparison to the dairy type Jamaica Hope based on the Jersey (Tanner Davy, 1869; Mason, 1996: Duffus and Jennings, 2009).

Devons were imported to Canada before 1865 but there was very little interest shown subsequently. In 1970, however, three bulls and 12 cows/heifers were imported. All three males but only five females were registered in the General Stud and Herd Book (which is the proxy registration authority for species and breeds that do not have a society of their own) maintained by the Canadian Livestock Records Corporation. Two of the bulls and the five females were registered to two herds in Alberta. The third bull was first registered at the University of Manitoba before being transferred in 1977 to one of the herds that had imported the other animals. Only six offspring in total were registered for the three bulls but several offspring were registered for each of the cows. The General Herd Book has had a total of 193 pure and 45 grade Devons registered. Devons were used in Canada to sire crossbred offspring on extensive rangelands with only a small nucleus being maintained pure for replacements. There have been no registrations of Devons since 1981 but a further four polled cows were imported in 1987[8].

A total of 70 bulls and 38 cows and heifers was exported to Rhodesia, mostly prior to 1923 when the country was governed by the British South Africa Company (BSAC). Only five bulls and four cows and heifers were imported after the introduction of "Responsible Government" in 1923, the last group of one bull and four females arriving in 1948. In addition to "southern" Rhodesia, five bulls were imported to Northern Rhodesia in 1913 which was the year that the BSAC territory imported its greatest number of animals (28 bulls and 7 females) that included two bulls purchased by the Leibig Extract of Meat Company ("oxo"). Perhaps this last large increase was in response to a lament by the home Society's President in 1911 that, although there had been great success in South Africa, this had not been followed up and Rhodesia had taken a consignment of Shorthorns and Sussex cattle for working purposes (DCBS, 1911 and other volumes). The upsurge in Devon imports was, however, short-lived and it was eventually the Sussex that became the most popular British breed in Rhodesia before it in turn was overtaken by zebu cattle (personal observations by the author, 1965-2000).

---

[8] Information for this paragraph was very kindly provided by Ronald Black, General Manager in Ottawa of the Canadian Livestock Records Corporation.

In the twentyfirst century semen has been sent to Saint Helena (in 2005) and live animals to France (2008) (author's personal knowledge).

Conclusion

In the nine hundred plus years since Domesday the Devon, from being a prized beast, became a prized beast. In its first guise its worth and weight were at the front and adapted to the yoke. In its second its value and virtue had shifted to the rear in order to satisfy the butcher's slab. The transition from Model 1 to Model 2 took considerable time but was probably most rapid, and contemporaneous with Bakewell's improvement of the Longhorn and the Colwill Brothers of the Shorthorn, in the second half of the eighteenth century. The reasons for the shift were several. Oxen were being partially replaced by the faster (and by then more powerful) horse for agricultural work. There was consumer demand for beef of better quality than that derived from 6-8 year old recycled bullocks. Most important, however, was the combination of the beginnings of an understanding of nutrition, changes in agricultural practices that allowed clover and other legumes into the rotation and that provided other improved feeds – turnips, "cabbages" (kale and kohl rabi and in Devon especially a very large, firm hearted, blue-tinged cabbage known as a "flatpoll") and oilseed cakes – and the realization that improved nutrition and feeding was better used on a younger beast that would grow faster and flesh better than wasting this technology on a superannuated ox (Trow-Smith, 1959).

In spite of its new beef qualities many early importing countries chose Devons for their still undoubted draught power. These included the USA, New Zealand, South Africa and Rhodesia: in the case of the last, when it imported Devons in the early 1900s, it was still recovering from the loss of more than 90 per cent of its work oxen as a result of the rinderpest pandemic of the late 1890s (Mack, 1970: Wilson, 2007). Exports to most countries since 1850 have, however, been for beef production as a pure bred animal, as a crossing sire or in at least three cases for the creation of new breeds.

Just as they were in the seventeenth century, the twentieth century Devons did "not grow near as large as the Dutch cattle, do not give as much milk, and are much cheaper; but they fat and tallow well" (Van der Donck, 1656: Jameson, 1909). In the 1970s/1980s consumers were led to believe their tastes had changed and they needed to eat meat without external or marble fat. The new supermarket chains demanded bigger animals and the smaller traditional

British breeds faded with the onslaught of the continental types sought by the wholesale buyers. In response to this challenge the modern Devon has emerged from its own version of global recession and has become a larger, longer and leaner beast (Figure 6). In so doing it has not lost the qualities for which it has long been renowned – docility, fertility, ease of calving, thriftiness and the quality of its meat – and now competes with other major beef breeds, be they British or "continental".

The Devon both at home and overseas is associated with its ability to produce top quality beef as a pure bred and as a cross under varying climatic conditions. In a period when several other breeds resorted to using genetics of various backgrounds to "improve " (and to modify irredeemably) their animals the Devon has used this approach to genetic progress with caution and has made considerably more use of within-breed selection. This has been more the case abroad than at home, however, where it is often the proud boast of the Breed Societies that Devons retain the genetic purity that was developed over several centuries in their original homeland.

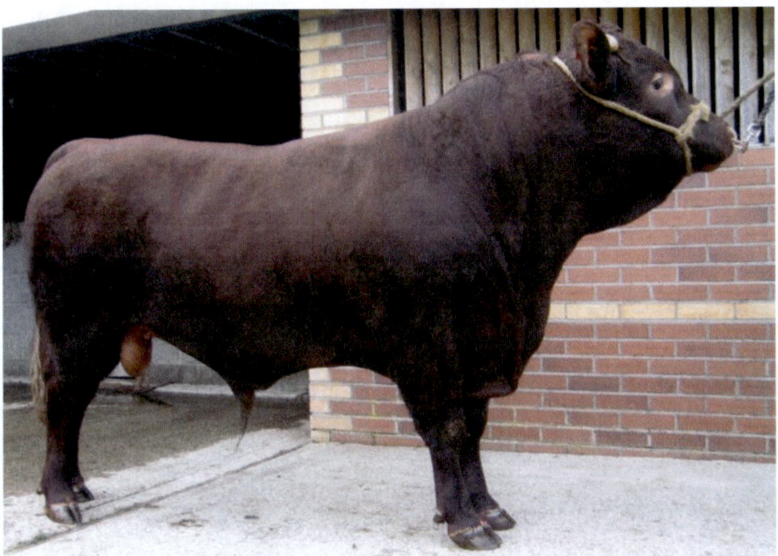

Bartridge China Boy 8th M08044, 18 month (556 days) weight 668 kg.

Figure 6. A modern Devon yearling bull.

This does not mean that the breed has not continued to develop. Progressive management and breeding techniques have seen Devons adapt well to the modern beef industry worldwide. Performance recording has identified superior animals and polled strains are now widely available.

Devons now have a broad genetic diversity and a range of maturity patterns that enable them to perform in a wide variety of economic and environmental situations. The features prominent in selection criteria for the new Devon were good muscle development, a docile temperament and the ability to thrive under adverse conditions. Success in these ideals is the foundation of the Devon of today and they are qualities of great value to the modern commercial beef and dairy industries. Devons have been proved to be one of the most efficient breeds in numerous controlled trials as well as under commercial conditions. In the existing reality of changing weather patterns worldwide, an economic climate of increased management and feed costs and unfavourable input-output ratios, efficiency is paramount. Today's consumers are demanding that the beef they eat is more natural, produced from grass, has not been tainted with growth promoters or truck loads of veterinary pharmaceuticals, is better tasting and has generally caused little damage to the environment in its production. If sustainable farming is to continue it will be the hardy most efficient breeds that fulfil environmental and economic criteria and Devons are amongst those best placed to satisfy these demands.

The Devon is not a historic curiosity. It is a practical weather-proof animal that requires the minimum of husbandry and is capable of finding its sustenance in distinctly inhospitable surroundings. The breed has a proud and productive past that will sustain it in the future.

Acknowledgments

Albert Beer, formerly Secretary of the Devon Cattle Breeders' Society, reviewed this manuscript and made valuable comments for its improvement. I thank Charl Hunlun of the South African Studbook and Ron Black of the Canadian Livestock Records Corporation for information on Devons in their respective countries. John H Thomas, Angus and Joy Cottey and Kate Palmer and Robert James allowed access to their collections of Davy's Herd Books.

REFERENCES

ADCBA. (2009). American Devon Cattle Breeders Association. www.americandevon.com/devon-cattle-articles/devon-cattle-history; americandevon.com/about/adca-percentage-registry-program.

AMDCA. (2009). American Milking Devon Cattle Association. http://www.milkingdevons.org.; American Livestock Breeds Conservancy, http//:www.albc-usa.org.

Beer, A. (2009). Farming Talk. *Exmoor the Country Magazine*, Issue No 46 Spring 2009, 10-11.

Beer, A. and Beer, S. (1998). Red Rubies - Jewels in Exmoor's Crown. *Exmoor the Country Magazine*, Issue No 4 Autumn 1998, 16-18.

BOAF. (1912). Agricultural output of Great Britain (Census of Production Act, 1906). London, UK: Board of Agriculture and Fisheries CD 6277, 1912 37.

Bowling, G.A. (1942). The introduction of cattle into colonial North America. *Journal of Dairy Science*, 25, 129-154 (West Virginia Agricultural Experiment Station Scientific Paper No. 277).

Brinkman, L.W. (1964). The Historical Geography of Improved Cattle in the United States to 1870 (unpublished. Ph.D Dissertation). Madison, WI: University of Wisconsin.

Cirne-Lima, E.O. (2008). Report from the Brazilian Devon Breeders Society's President, 8th World Devon Conference and Red Ruby Devon Tour, Australia, 2008. http://www.americandevon.com/devon-cattle-articles/brazilian-world-conference-report.

DCBS. (1911). Davy's Devon Herd Book Volume 34. Iddesleigh Devon UK: Devon Cattle Breeders' Society.

DCBS. (1917). Davy's Devon Herd Book Volume 40. Iddesleigh Devon UK: Devon Cattle Breeders' Society.

DCBS. (1924). Davy's Devon Herd Book Volume 44. Iddesleigh Devon UK: Devon Cattle Breeders' Society.

DCBS. (2009). Davy's Devon Herd Book Volume 132. Iddesleigh Devon UK: Devon Cattle Breeders' Society.

DCBSA. (2009). History of Devon cattle in Australia. Gloucester, NSW, Australia: Devon Cattle Breeders Society of Australia. http://www.devoncattle.com/history-australia.php.

Deane, C. (ed). (1851). History of Plymouth Plantation by William Bradford, Second Governor of the Colony, (Reprinted from the Massachusetts Historical Collections). Boston, MA: Privately Printed.

DEFRA. (2006). Agricultural land use in England and the South West 2006, Department of the Environment, Food and Rural Affairs, http://www.defra.gov.uk/esg/work_htm/publications/cs/farmstats_web/2_SURVEY_DATA_SEARCH/COMPLETE_DATASETS/regional_level_datasets.htm.

Dmitriev, N.G. and Ernst, L.K. (1989). Animal genetic resources of the USSR (Animal Production and Health Paper No 65). Rome, Italy: Food and Agriculture Organization.

Duffus, B. and and Jennings, P. (2009). The Current State of the Jamaica Cattle Sector. Kingston, Jamaica: Jamaica Livestock Association Limited.

DVSAH. (1926). Dare es Salaam, Tanganyika: Government Printer.

DVSAH. (1927). Dare es Salaam, Tanganyika: Government Printer.

DVSAH. (1928). Dare es Salaam, Tanganyika: Government Printer.

Finberg, H.P.R. (1951). Tavistock Abbey: a study in the social and economic history of Devon. Cambridge, UK: Cambridge University Press. p132.

Hall, S.J.G. and Clutton-Brook, J. (1989). Two hundred years of British farm livestock: Devon. London, UK: British Museum (Natural History). pp67-69.

Hamilton, A.H.A. (1878). Quarter Sessions from Queen Elizabeth to Queen Anne. London, UK: Sampson Low. p81.

Jameson, J.F. (ed). (1909). Narratives of New Nederland, 1609-1664 (Original Narratives of Early American History). New York, NY: Charles Scribner's Sons.

Kelley, R.B. (1959). Native and adapted cattle. Sydney, Australia: Angus and Robertson.

Mack, R. (1970). The great African cattle plague epidemic of the 1890s, *Tropical Animal Health and Production*, 2, 210-219.

Mason, I.L. (1996). A World Dictionary of Livestock Breeds, Types and Varieties (Fourth Edition). Wallingford UK: CAB International. p35.

MOAF. (1921). Census of Pedigree Live Stock. London, UK: Ministry of Agriculture and Fisheries.

Morgan, F.W. (1940). The Domesday Geography of Devon. Transactions of the Devonshire Association for the Advancement of Science, Literature and Art, 72, 305-331.

NZRDCBA. (2009). Devon history in New Zealand. Fielding, New Zealand: New Zealand Red Devon Cattle Breeders Association. http//:www.reddevoncattle.co.nz/information/cattle-history-i-6.html

RBST. (2009). The Ark, the Magazine of the Rare Breeds Survival Trust, Spring 2009.

Risdon, T. (1957), The Chorographical Description or Survey of the County of Devon, 1811 (quoted in Trow-Smith, R. (1957)).

SASB. (1971). The Devon Breed in South Africa. Bloemfontein, South African: South African Stud Book Association Jubilee Edition 21, 41-42.

Scott Watson, J.A. and Hobbs, M.E. (1951). Great Farmers (Revised Edition) London, UK: Faber and Faber.

Sinclair J. (1893). History of the Devon Breed of Cattle. London, UK: Vinton.

Splan, R.K. and Sponenberg, D.P. (2004). Characterization and conservation of the American Milking Devon. *Animal Genetic Resources Information*, 34, 11-16.

Tanner Davy, J. (1869). A Short History of the Rise and Progress of the Devon Breed of Cattle. *Journal of the Royal Agricultural Society of England*, 30, 107-130.

Thornton, C. (1993). Red Rubies: A History of the Devon Breed of Cattle. Manchester, UK. *Gabriel Communications*. pp23-25.

Trow-Smith, R. (1957). A History of British Livestock Husbandry, to 1700. London, UK: Routledge and Kegan Paul.

Trow-Smith, R. (1959). A History of British Livestock Husbandry, 1700-1900. London, UK. Routledge and Kegan Paul.

US Census. (1920). V Agriculture: General Report and Analytical Tables, pp. 640-643. Washington, DC: US Census Bureau.

Van der Donck, A. (1656). Description of the New Netherlands, 1656. New York Historical Society Collections Series 2, 1, 165.

Walton, J.R. (1999). Pedigree and productivity in the British and North American cattle kingdoms before 1930. *Journal of Historical Geography*, 25, 441-462.

Wilson, R.T. (2007). The one-humped camel in Southern Africa: their presence and use in Rhodesia (Zimbabwe) in the early twentieth century. *South African Animal Science*, 8, 21-28.

Wilson, R.T. (2011). The Indigenous Domestic Livestock of the County of Devon, England. *Animal Genetic Resources*, xx, yyy-zzz.

In: Native Species
Editors: L. Marin and D. Kovač
ISBN: 978-1-61470-613-7

Chapter 3

NATIVE PLANTS OF NORTHEASTERN MEXICO: THEIR POTENTIAL IN THE RESTORATION OF DAMAGED ECOSYSTEMS

Rahim Foroughbakhch,[1*] Jorge Luis Hernández Piñero,[1] Marco A. Alvarado Vázquez,[1] Artemio Carrillo Parra,[2] Carlos Gerardo Velasco Macías,[3] and Alejandra Rocha Estrada[1]

[1]Departamento de Botánica, Facultad de Ciencias Biológicas, Universidad Autónoma de Nuevo León, Mexico
[2]Facultad de Ciencias Forestales, UANL, Linares, Nuevo León, Mexico
[3]Areas Naturales Protegidas, Parques y Vida Silvestre de Nuevo León, Mexico

ABSTRACT

Native plants naturally occur in the region in which they evolved and adapted to the physical and biotic processes characteristic of that region,

* Corresponding author: Facultad de Ciencias Biológicas, UANL, A.P. F-2, 66451 San Nicolás de los Garza, N.L., México. E-mail: rahim.f@gmail.com.

e.g. climate, soil, time of rainfall, drought, and frost as well as interactions with other species inhabiting the local community. Thus native plants possess certain traits that make them uniquely adapted to local conditions providing a practical and ecologically valuable alternative for landscaping, conservation and restoration of degraded ecosystems.

Mexico's ecosystem is a vast one as the country covers a wide variety of terrains, soils and climates with different geography characteristics which favor the development of native species. For instance, in those arid and semiarid regions of northeastern Mexico there is an extensive native plant list of more than 3,100 species.

Regions of the coastal plain of the Gulf of Mexico are dominated by a dense diversity of native plants. The vascular flora comprises 159 families, 1034 genera, 3181 species and 109 intraspecific taxa. The most diverse families are: Asteraceae (148 genera and 481 species), Poaceae (97 genera and 301 species), Fabaceae (76 genera and 258 species), Cactaceae (37 genera and 139 species) and Euphorbiaceae (15 genera and 85 species) among others. Native plant inventory in northeastern Mexico is mostly composed of deciduous and evergreen or perennial species which are characterized by a wide range of growth patterns, diversity in leaf longevity, dynamics of growth and contrasting phenological developments. This resource is used by the inhabitants of the region in various ways such as forage for livestock, coal, timber, shelving, food, herbal, medicine and seed supply for reforestation and ecological restoration. Additionally, the native plant provides a habitat for wildlife and a cover layer that prevent soil erosion. For these reasons native plants develop various morphological and physiological characteristics suitable for adaptation to adverse environmental factors, particularly drought stress and high temperatures.

Numerous vegetal species account for the bulk of the bush vegetation providing a characteristic landscape. The most important of these plants consist of various species of *Acacia* including *A. rigidula*, *A. farnesiana*, *A. berlandierii*, *A. smallii* and *A. tortuosa*, and some other important species such as *Prosopis laevigata*, *P. glandulosa*, *Leucaena leucocephala*, *Celtis pallida* (desert hackberry), *Condalia ericoides* (javelina bush), *Leucophyllum frutescens* (cenizo or sometimes called purple sage), and *Aloysia wrightii* (common bee-brush or white brush). These species favor the retention of soil due to their high sprouting potential, ease of regeneration after frost and forest fires in addition to being accepted by the people so they may be used successfully to restore degraded areas and initiate programs aimed to the protection and conservation of forest elements in northeastern Mexico.

INTRODUCTION

According to the Science and Policy Working Group of the Society for Ecological Restoration International (2004) the ecological restoration is an intentional activity that initiates or accelerates the recovery of an ecosystem with respect to its health, integrity and sustainability. Frequently, those ecosystems requiring restoration have been degraded, damaged, transformed or entirely destroyed as the direct or indirect result of human activities. In some cases, these impacts to ecosystems have been caused or aggravated by natural agents such as wildfire, floods, storms, or volcanic eruptions, to the point at which the ecosystem cannot recover its predisturbance state or its historic developmental trajectory.

On the other hand, the term restoration has been given to any ecological process aimed to recovering the environmental conditions prevailing at a given site which were adversely affected for whatever reason. The first ecological restoration actions reported in the literature with this particular approach were carried out in grasslands of the state of Wisconsin, USA (25 hectares) by Aldo Leopold, in 1935, who is recognized as one of the pioneers in the field (Sol-Sánchez et al., 2002).

The term restoration is still not well understood and has been incorrectly applied to various activities of reforestation or natural regeneration undergoing similar concepts but different methodology and aims pursued in each case. However, restoration comprise broad applications to the rescue of several areas affected by natural (hurricanes, thunderstorms, fires, floods, landslides, etc.) or anthropogenic causes (pollution, deforestation, burning, etc.) and for species that are in some degree of vulnerability.

In a relatively short period of time the vegetation of Mexico has suffered extensive human disturbances. Very few areas of the country remain with undisturbed ecological communities. The footprint of deforestation, bush burning, overgrazing and their consequences on vegetation and fertility of soil are noticeable in almost any country's landscape. Faced with such a serious impact on rural productivity and biodiversity it becomes as an urgent priority to begin to develop conservation procedures to reverse the terrible decline in an intelligent way. One of the plant communities mostly impacted by clearing have been the tropical forests and to a lesser extent the mountain forests and xerophilous shrublands. In addition to this, fires, illegal felling and exploitation of valuable species have contributed to the decline of forest cover in many countries.

The gradual conversion of Mexico's forests on islands of vegetation has resulted in the destruction of some habitats and pollution of others. Biological corridors, which in the past formed a continuous passage for wildlife in the mountains and coastal plains of the Gulf of Mexico and the Pacific Ocean are now fragmented. Many species of animals and plants remain isolated and therefore their chances of mating, genetic exchange and evolution have been lowered due to the reduction of their natural variation and ecological niches.

Within the diversity of plant communities of our country, bush vegetation is of particular interest because it has an intrinsic ability to easily recover from impacts. This is why this type of vegetation can be easily regenerated in a short time after adverse natural conditions such as drought, frost, or induced factors like fire or overuse, providing a vegetative cover to reduce soil loss and promotes the development of other species.

However, there is scarce or no information about the mechanism of reaction of scrub vegetation after natural or induced impacts neither about the use of shrub species in ecological restoration (Sendy, 1987).

Ecosystem Degradation and Ecological Restoration

Environmental Services of Ecosystems

In general terms, ecosystems provide a wide variety of goods and services locally, nationally and globally. Environmental goods means those products (wood, fruits, seeds, bark, leaves, etc.) that are used by humans for consumption or marketing, while environmental services are referred to the ability of ecosystems to generate products useful to man, including the regulation of gases (oxygen production and carbon sequestration), scenic beauty, biodiversity protection, soil and water among others. The environmental services are divided into three types:

a. Services provision: production of food, timber, medicines and water.
b. Regulatory services, biodiversity conservation, carbon sequestration, climate balance, control of soil erosion, pests and natural disasters (like hurricanes), water provision and cleaning systems and genetic banks.

c. Cultural services, such as aesthetic landscape and the spiritual or religious value of some plant and animals species as well as recreation space.

Due to the transformation that mankind have been causing to ecosystems, these services are being lost or degraded which is why greater awareness and knowledge about the impacts of these changes is arising and protection schemes such as environmental service payments (ESP) have been developed. ESP are transactions through which land owners are paid by users of environmental services, which can be done both directly with the development of a market where users explicitly provide for the conservation and improvement of the service or through indirect ways usually mediated by the government through taxes or subsidies (Alba-Cuellar, 2008).

Degraded Ecosystems

It is considered that an ecosystem is degraded when the plant communities that dominate a given area are replaced with distant species which in turn drive it away from the corresponding climax community. The degradation of ecosystems brings along certain consequences such as the greenhouse effect, acid rain, destruction of the ozone layer, use of phosphates, visual or noise pollution as well as the increase in the volume of waste generated (Lamprecht , 1990). Other characteristics of degraded ecosystems are the considerable reduction in diversity, productivity and habitat. Degraded terrestrial ecosystems are characterized by loss of vegetation and soil. Likewise, aquatic ecosystems are often characterized by its polluted waters that few species can tolerate. Specifically, degraded ecosystems are those where productivity and diversity has been reduced in such a way that it is unlikely to recover its original condition unless special rehabilitation actions are applied (IUCN, UNEP and WWF, 1991).

Ecological Restoration

Ecological restoration is the process by which a site is intentionally altered to re-establish an ecosystem. To reach this goal, the structure, function, diversity and dynamics of the ecosystem to restore is emulated. On the other hand it has been stated that ecological restoration is an emerging science with a profound importance in conservation biology (Kondolf, 1995, Kondolf and Micheli, 1995).

There are three different positions on the meaning of ecological restoration. A fundamentalist view consists in considering restoration as a returning back to the original conditions of the occurring natural communities of a region, including its original biodiversity, achieving a certain degree of stability again without further management. Return to the original situation may still be possible in places such as disturbed areas of nature reserves and protected areas where only a part of the original community has been altered.

In contrast, in many places a more practical second option will be only possible to apply combined with productive activities. In this case, ecological restoration would be aimed to the recovering of the main environmental functions of the original ecosystem that will maintain stability in fertility, soil conservation and hydrological cycle even though some diversity could be lost, the stability of system needed to be handled and some exotic species could be introduced to the area.

Landscape restoration is the third way to envision restoration. In this case the goal is the development of a nice-looking and friendly landscape over a site with unpleasant characteristics, for example, a sanitary landfill.

The level of destruction present on the vegetation cover, soil fertility and the regeneration capacity of the native vegetation of the affected sites will set the pace of the origin and biological characteristics of those species that may be used for any restoration strategy. Places with a relatively mild level of deterioration could retain their natural mechanisms of regeneration such as the presence of an edaphic bank of seeds and alive vegetative structures, seed-rain and still fertile soils. A greater level of deterioration may require manipulations that include the environmental improvement of the site through the use of plants which enhance the quality of the soil and microclimate combined with the reactivation of seed-rain from nearby areas which were conserved through biological ways. Finally, heavily disturbed areas or regions, where a natural or induced invasion of alien plants occurred, could require more drastic actions including the total removal of the invasive vegetation, environmental improvement of the site by local or adequate introduced plant species, and may even require changes with landscape engineering techniques to improve the conditions for the successful establishment of the plants used for reforestation.

There is also a so high degradation level in which no recovery would take place or it would occur too slowly (Finegan, 1993 and Machlis, 1993 in Gálvez, 2002) due to the surpassing of whichever ecological limit. Examples of such limits are:

- When the area is vulnerable to recurrent disturbances (e.g. fire).
- When there are few remnants of the original communities and the dispersal distances are long (e.g. islands).
- When the rate of dispersion of major animal or plant species is low.
- When certain key species cannot recolonize without outside help.
- When there is an excessive amount of weeds and pests.

The Restoration Process

The Society for Ecological Restoration International Science and Policy Working Group (2004) states that the restoration process attempts to return an ecosystem to its historic trajectory. Historic conditions are therefore the ideal starting point for restoration design. The restored ecosystem will not necessarily recover its former state since contemporary constraints and conditions may cause it to develop along an altered trajectory. The historic trajectory of a severely impacted ecosystem may be difficult or impossible to determine with accuracy. Nevertheless, the general direction and boundaries of that trajectory can be established through a combination of knowledge of the damaged ecosystem's pre-existing structure, composition and functioning, studies on comparable intact ecosystems, information about regional environmental conditions, and analysis of other ecological, cultural and historical reference information. These combined sources allow the historic trajectory or reference conditions to be charted from baseline ecological data and predictive models, and its emulation in the restoration process should aid in piloting the ecosystem towards improved health and integrity.

Restoration represents an indefinitely long-term commitment of land and resources, and a proposal to restore an ecosystem requires thoughtful deliberation. Collective decisions are more likely to be honored and implemented than are those that are made unilaterally. For that reason, it behooves all stakeholders to arrive at the decision to initiate a restoration project by consensus. Once the decision to restore is made, the project requires careful and systematic planning and a monitored approach towards ecosystem recovery. The need for planning intensifies when the unit of restoration is a complex landscape of contiguous ecosystems.

Interventions employed in restoration vary widely among projects, depending on the extent and duration of past disturbances, cultural conditions that have shaped the landscape, and contemporary constraints and opportunities. In the simplest circumstances, restoration consists of removing or modifying a specific disturbance, thereby allowing ecological processes to bring about an independent recovery. For example, removing a dam allows the return of an historical flooding regime. In more complex circumstances, restoration may also require the deliberate reintroduction of native species that have been lost, and the elimination or control of harmful, invasive exotic species to the greatest practicable extent.

When the desired trajectory is realized, the ecosystem under manipulation may no longer require external assistance to ensure its future health and integrity, in which case restoration can be considered complete. Nevertheless, the restored ecosystem often requires continuing management to counteract the invasion of opportunist species, the impacts of various human activities, climate change, and other unforeseeable events. In this respect, a restored ecosystem is no different from an undamaged ecosystem of the same kind, and both are likely to require some level of ecosystem management. Although ecosystem restoration and ecosystem management form a continuum and often employ similar sorts of intervention, ecological restoration aims at assisting or initiating recovery, whereas ecosystem management is intended to guarantee the continued well-being of the restored ecosystem thereafter.

Sol-Sánchez et al. (2002) proposed a model for ecological restoration of disturbed areas, which consists of 3 stages: 1) Identification of the area to be restored, 2) Definition of the restoration strategy, 3) Establishment of restoration plots, 4) Maintenance and periodic evaluation of the restoration plots and 5) Field data and restoration indicators.

For its part, Machlis (1993) mentioned that there are three basic ways to restore a degraded area: 1) Recover: re-plant land cover with appropriate species, 2) Rehabilitate: using a mixture of native and exotic species to recover the area, 3) Restore: re-establishing in the place the original set of plants and animals with approximately the same previous population.

Here are other useful methods in ecological restoration, such as agroforestry systems and tree domestication.

Agro-Silvopastoral Systems

One way that we can use domesticated wild plants in combination with agricultural production is through agrosilvopastoral or agroforestry systems which may present a variable spatial structure depending on the type of ecosystem in question and the intended purpose in its design. These are combinations of cultivated trees and shrubs with crops or pasture which vary in timing, species composition, spatial arrangement and biological interactions that occur between their components. If these systems are well designed, wood components can provide valuable services such as shade, farming protection against wind, control of eolic and water erosion, reduction of evapotranspiration, accumulation of organic matter in soil, atmospheric nitrogen fixation in soil and vegetation, efficient recycling of mineral nutrients, water retention and infiltration into the soil and habitat for native species, while supplementary products originating from cultivation or livestock occurs on the site at the same time. These products include: fodder, firewood and charcoal, poles, green fertilizers, edible beans, honey, timber, fruits, dyes, fibers, among other things. The other benefits provided to us by agrosilvopastoral systems are essentially improved soil conservation and the reduced use of fertilizers. Unfortunately, pressure of economic or social structures derived from the forms of land tenure make these systems difficult to implement.

Legumes (Fabaceae) play a major role in tropical agroforestry systems for their ability to fix atmospheric molecular nitrogen into organic molecules, however best results are achieved with a combination of trees with leaves containing different levels of carbon and nitrogen concentration, as this promotes a better quality of the humus formed and therefore, over time, improved cation exchange capacity and improve the edaphic structure of soil conglomerates whose shape, size and arrangement determine the erosion and fertility degree.

Domestication of Trees

Whether it is about ecological restoration, reforestation or agroforestry systems development, domestication is the first stage of tree management.

Here we use the word domestication in a very different sense than that it is used in ethnobotany. The general procedure for tree domestication is as follows: 1) Adoption of native or exotic plants by looking at the existing flora species capable to provide the desired service. 2) Identify and collect the obtainable germplasm. 3) Search phenotypes optimal for the intended purpose. 4) Otimize the methods of propagation. 5) Improve the valuable qualities of trees.

Valuable plants for restoration and reforestation should have the following quality characteristics:

- Easy to propagate.
- Tolerant to limiting conditions such as low fertility, drought, compacted soils, high or low pH, salinity, and so on.
- Rapid growth and good production of organic matter such as compost from fallen leaves, preferably with a high ratio of C/N.
- Some added-value to its restorative effect, such as fuelwood, charcoal, fodder, nutritious, edible pods, wood, or nectar.
- No tendency to acquire an uncontrollable weed invasive feature.
- Presence of nitrogen fixing nodules or mycorrhizae to compensate for the low level of nitrogen, phosphorus and other nutrients in the soil.
- Tendency to favor the re-establishment of the flora and fauna populations by providing food and habitat. Over time, clusters or patches of these plants may grow naturally to facilitate nutrient recycling, fertility preservation, and gradually approaching the general natural way of community growth.

Attributes of Restored Ecosystems (Society for Ecological Restoration International Science and Policy Working Group, 2004)

An ecosystem has recovered - and restored - when it contains sufficient biotic and abiotic resources to continue its development without further assistance or subsidy. It will sustain itself structurally and functionally. It will demonstrate resilience to normal ranges of environmental stress and disturbance. It will interact with contiguous ecosystems in terms of biotic and abiotic flows and cultural interactions.

The nine attributes listed below provide a basis for determining when restoration has been accomplished. The full expression of all of these attributes is not essential to demonstrate restoration. Instead, it is only necessary for these attributes to demonstrate an appropriate trajectory of ecosystem development towards the intended goals or reference. Some attributes are readily measured. Others must be assessed indirectly, including most ecosystem functions, which cannot be ascertained without research efforts that exceed the capabilities and budgets of most restoration projects.

1. The restored ecosystem contains a characteristic assemblage of the species that occur in the reference ecosystem and that provide appropriate community structure.
2. The restored ecosystem consists of native species to the greatest practicable extent. In restored cultural ecosystems, allowances can be made for exotic domesticated species and for non-invasive ruderal and segetal species that presumably co-evolved with them. Ruderals are plants that colonize disturbed sites, whereas segetals typically grow intermixed with crop species.
3. All functional groups necessary for the continued development and/or stability of the restored ecosystem are represented or, if they are not, the missing groups have the potential to colonize by natural means.
4. The physical environment of the restored ecosystem is capable of sustaining reproducing populations of the species necessary for its continued stability or development along the desired trajectory.
5. The restored ecosystem apparently functions normally for its ecological stage of development, and signs of dysfunction are absent.
6. The restored ecosystem is suitably integrated into a larger ecological matrix or landscape, with which it interacts through abiotic and biotic flows and exchanges.
7. Potential threats to the health and integrity of the restored ecosystem from the surrounding landscape have been eliminated or reduced as much as possible.
8. The restored ecosystem is sufficiently resilient to endure the normal periodic stress events in the local environment that serve to maintain the integrity of the ecosystem.
9. The restored ecosystem is self-sustaining to the same degree as its reference ecosystem, and has the potential to persist indefinitely under existing environmental conditions. Nevertheless, aspects of its

> biodiversity, structure and functioning may change as part of normal ecosystem development, and may fluctuate in response to normal periodic stress and occasional disturbance events of greater consequence. As in any intact ecosystem, the species composition and other attributes of a restored ecosystem may evolve as environmental conditions change.

Other attributes gain relevance and should be added to this list if they are identified as goals of the restoration project. For example, one of the goals of restoration might be to provide specified natural goods and services for social benefit in a sustainable manner. In this respect, the restored ecosystem serves as natural capital for the accrual of these goods and services. Another goal might be for the restored ecosystem to provide habitat for rare species or to harbor a diverse genepool for selected species. Other possible goals of restoration might include the provision of aesthetic amenities or the accommodation of activities of social consequence, such as the strengthening of a community through the participation of individuals in a restoration project.

Native Plants and Ecological Restoration

Native species are those that occur in the region in which they evolved (Richards et al. 1998). Plants evolved over geologic time in response to physical and biotic processes characteristic of a region: the climate, soils, timing of rainfall, drought, frost, and interactions with the other species inhabiting the local community. Thus native plants have acquired certain traits that make them uniquely adapted to local conditions providing a practical and ecologically valuable alternative for landscaping, conservation and restoration of degraded ecosystems and as livestock forage.

As the public becomes more concerned about the environment the interest in the preservation and restoration of native plant communities increases as well. Native plants are valued for their economic, ecological, genetic, and aesthetic benefits in addition to the growing societal belief about their intrinsic value as living species.

Using native plants to restore the landscape or as substitute for exotic ornamental planting can help to reverse the trend of species loss. Although the methods may differ, native plants require the same level of care in establishment as do ornamental plants. If the environment has been altered significantly through human activities, some work will be necessary to recreate a friendlier environment to native plants. However, in the long run, native plants will form self-sustaining plant communities in most cases that do not require maintenance in a large extent because they are adapted to the local region, tend to tolerate damage from freezing, drought, common diseases, and predation (Brown and Lugo, 1990).

Native plant species provide the keystone elements for ecosystem restoration increasing the local population so providing numerous benefits. There are specific associations of mycorrhizae with plants, invertebrates with woody debris, pollinators with flowers, and birds with structural habitat that can only be rebuilt by planting native plants (Lieberg, 1993).

Advantages of Native Plants

The uses of native plants in restoration strategies are numerous. Native plants add beauty to the landscape and preserve our natural heritage, provide food and habitat for native wildlife, serve as an important genetic resource for future food crops or other plant-derived products, help slow down the spread of fire by staying greener for a longer time, decrease the amount of water needed for landscape maintenance, require very little long-term maintenance if they are properly planted and established, produce long root systems to hold soil in place, and protect water quality by controlling soil erosion and moderating floods and droughts among others.

Diversity of Plant Communities in Northeastern Mexico

In the coastal plains of the northern Gulf of Mexico the vegetation type characteristic is the submontane scrub and the Tamaulipan thorn scrub. There are also some oak forests, and pine and oak-pine forests at the higher places. In the physiographic area of the Sierra Madre Oriental a microphile desert scrub and desert rosette scrub currently develop. The most common species are gobernadora, hojasén, mesquite and cactus.

In other areas of the Sierra Madre Oriental there is diverse vegetation although forests and scrublands mainly occur. The first is dominated by pine and the second by rosette and submontane desert scrub and by chaparral. Other types of vegetation are natural and introduced grasslands, halophytic plants and moorland vegetation.

In more than half of the land there is vegetation suitable for the exploitation of forest resources for local inhabitant's subsistence and other areas which can be exploited for industrial purposes.

FORESTS

Pine

The distribution of these forests may occur in temperate and humid places with the heat factor decreasing as going up in altitude; it is located from 900 to 1,000 masl. Pinus species are commonly associated with some species of oak (*Quercus* spp.) and madrone trees (*Arbutus* spp.).

Characteristic pine species are *Pinus teocote, P. pseudostrobus, P. arizonica, P. ayacahuite,*and *P. hartwegii*. We may find variations of these forests such as those of *P. cembroides*. They grow in areas of low rainfall between 2,000 to 2,600 masl on shallow rocky soils. This type of forest has open spaces of short trees 4 to 8 m high with rounded cups and trunks with diameters of 30 to 40 cm at breast height. The main arboreal species in this type of forest are *P. cembroides, P. greggii, Juniperus flaccida, J. deppeana* and *Arbutus xalapensis*. It is also found along the edges of the scrub desert plateau and often associated with species such as *Yucca carnerosana*, *Yucca filifera, Agave* spp and various shrubs.

Coniferous Forest

The hallarin-oyamel (*Pseudotsuga-Abies*) forests are a variant of the temperate forests characterized by species of great scenic attraction with typical pyramidal shape trees. This type of vegetation consists of tall trees 15 to 25 m which are located in canyons covered with cold and wet weather, usually between 2,000 to 2,500 meters. The species that dominate in canopy are *Pinus greggii* and *Pinus hartwegii* along with *Pseudotsuga rnenziesii* and *flahaulti*, *Abies vejari*, and *Cupressus arizonica*.

There is another type of forest in some small restricted areas forming patches of cedar *Cupressus* spp. forests. These trees are large and reach 10 to 15 m high, a characteristic species is the *Cupresus arizonica*.

Another community is the forest of *Juniperus* spp. which is associated with forests of pines and oaks. It grows as a tree or shrub depending on environmental conditions. It is known that these species of trees develop very slowly in rocky limestone soil with low humidity. The most common species are *Juniperus monosperma* and *Juniperus flaccida*.

Pine-Oak Forest

This unit is comprised of forests of both broadleaf and needle leaf species all over from temperate to sub-humid areas. Despite the wide range of climatic conditions in the region the presence of this type of forest is due more to the thermal factor than moisture. This forest occurs between an altitude of 550 to 900 m mainly dominated *Quercus rysophylla*, *Quercus laeta*, *Quercus polymorpha* and *Quercus canbyi* frequently associated with *Pinus teocote* and *Pinus pseudostrobus.* Trees generally do not have a large size in diameter or height; these are 10 to 30 cm, and 10 to 14 m respectively.

Oak Forest

This type of forest is part of the temperate forest community over the area of the Sierra Madre Oriental in regions with soil from thin and rocky to places with deep and well drained soil. Physiognomically this forest type consist of trees and shrubs ranging from 12 to 15 meters high where its dominant genus is *Quercus*, closely linked ecologically and floristically with mixed and pine forests. The characteristic species are *Quercus rysophylla, Quercus laeta, Quercus polymorph, Quercus laceyi, Quercus canby, Quercus pungens, Quercus vaseyana* and *Quercus intricata*.

Scrub Lands

Submontane Scrub

It is a sub-arboreal shrubbery rich in life forms. The strength, size and distribution of the dominant and co-dependent species depend on the availability of water in the soil and subsoil. The dominant life forms are shrubs or trees from 1.2 to 1.8 m tall with small leaves, deciduous and subespine. They are located on the lower slopes of the mountains where actually form a large path that separates the elements of the desert scrub of the plains and the oak-pine forests developing in the upper slopes of the Sierra Madre Oriental.

Although it has morphological and ecological variants, in general terms in this scrub land grow the following most representative species: *Helietta parvifolia, Cordia boissieri, Sophora secundiflora, Gochnatia hypoleuca, Neopringlea integrifolia, Decatropis bicolor, Fraxinus greggii, Pithecellobium pallens*, *Leucophyllum frutescens, Acacia rigidula, Acacia berlandieri, Acacia farnesiana, Caesalpinia mexicana, Prosopis glandulosa, Dyospiros virginiana, Dyospiros texana* and *Cercidium macrum.* In some areas with protected habitats, abundant moisture and deep soils small clusters of *Quercus virginiana* are found.

Thorn Scrub and Mezquital

Natural plant communities and mesquite thorn scrub present physiognomic variants; species may be high or medium spiny subinerm, representing plant communities typical of the region of the Gulf Coastal Plain. Under favorable soil and moisture conditions the stems have well-defined trunks and tree forms are more than 6 meter. Dominant species for their abundance and coverage are *Prosopis laevigata, Ebenopsis ebano, Acacia rigidula, Castela texana, Celtis pallida, Cercidium macrum, Randia laetevirens, Cordia boissieri, Leucophyllum frutescens, Porlieria angustifolia, Opuntia leptocaulis, Opuntia engelmannii, Zanthoxylum fagara,* and *Bumelia celastrina,* and set apart from the rest *Yucca filifera* with up to 10 meter high. *Lupinus texensis* outstands in the herbaceous stratum.

Desert Scrub

This type of vegetation is characteristic of arid and semi-arid zones. Irregular distribution pattern is mainly due to the availability of soil moisture. This condition is another factor influencing the location and development of species of the desert scrub. Soils with texture from sandy to sandy-loam to clay have the ability to retain moisture and so support dense and tight communities while stony soils that retain little water are limited to support plant communities which are poor and very disperse. Two types of desert scrub are dominant:

Rosette desert scrub: This community occurs over mountain sides and slopes of various elevations where there are rocky outcrops and skeletal soils lithosols. The most conspicuous have succulent leaves grouped in rosettes, some with terminal spines or mucrones.

The most common species are Dasylirion texanurn, Hechtia glomerata, Berberis trifoliolata, Agave lechuguilla, Agave striata, Agave bracteosa, Euphorbia antisyphilitica, Opuntia leptocaulis, Echinocereus enneacanthus, Opuntia rnicrodasys, Echinocactus platyacanthus, Ferocactus pilosus and Opuntia spp.

Microphile desert scrub: This vegetation type is characterized by the dominance of species of bush type with small leaves or leaflets, often odorous. Cacti are found abundantly with spherical or flat stems. Palms like china palm or ixtle palm type are profusely located on flat terrains or in alluvial fans over hills or mountains in the highlands.

The characteristic species are Larrea tridentata, Flourensia cernua, Partheniurn incanurn, Fouquieria splendens, Mortonia greggii Partheniurn argentatum, Ephedra aspera, Viguiera stenoloba, Acacia rigidula, Castela texana, Acacia berlandieri, Celtis pallida, Prosopis glandulosa, Chilopsis linearis, Echinocactus platyacanthus, Yucca filifera, Yucca carnerosana, Opuntia rnicrodasys, Opuntia imbricata, Opuntia rastrera, Opuntia leptocaulis and Sporobolus airoides.

GRASSLANDS

This community is characterized by graminoid herbaceous species with thin, elongated leaves which can live with some other species of the families Asteraceae, Fabaceae and Chenopodiaceae. Natural climax grasslands occupy smaller areas in wide spaces within the desert scrub and, under specific soil conditions, in locations with poor drainage, flood, excessive salt or existence of gypsum.

Vegetation along Rivers and Streams

This community is closely linked to the orographic formation of the Sierra Madre Oriental as a large drainage basin with its tributary rivers and streams comprising both the arboreal and shrubby trees growing along their banks. These forests are composed of *Platanus occidentalis, Populus tremuloides, Salix nigra, Arundo donax, Adiantum capillus-veneris, Lobelia cardinalis,*

Ulmus crassifolia and *Taxodium mucronatum,* in association with abundant aquatic herbaceous or semi-aquatic, climbing and epiphytic species such as *Tillandsia paixtle*.

This kind of forest grows mainly in the river banks and over spacious flat terrains with surface drainage. These sites, dampened by perennial or lasting sporadic runoff allow species to acquire height and diameter dimensions much larger than those located in arid zones.

Annex 1 shows a list of native forest, desert, semi-arid and grassland communities typical of northeastern Mexico.

Importance of Vegetation in Northeastern Mexico

The plant associations found in northeastern Mexico are used in multiple ways. The wood of certain species are cut for firewood, charcoal and construction materials of houses and yards, some plants are collected for medicinal use while other species provide food to the rural population, wildlife and domestic livestock. Despite this forest structure of vegetation in northeastern Mexico the most important use given by the rural community is the extensive grazing for livestock. Finally, we should emphasize the ecological importance of soil retention and mitigation of ecological impacts (FAO-UNESCO, 1980).

Causes of the Deterioration of Vegetation in Northeastern Mexico

The disturbances that cause environmental degradation can be classified into natural and anthropogenic. Natural disturbances are forest fires, hurricanes, floods and landslides, while anthropogenic disturbances are intentional forest fires, logging, mining, clearing for agriculture, livestock use, urban construction, discharge of sediment, and disposal of pollutants (Figures 1, 2 and 3).

The different types of disturbances can also be classified according to the size of the impacted area: a) large disturbances (over 10 hectares), b) medium (1 to 10 hectares) and c) small (less than one hectare). Depending on the intensity or damage caused to ecosystems, they are classified as serious or severe, medium and mild.

Characteristics of the Disturbed Areas

A disturbance area is one that has lost all or part of its attributes or, in other words, its function (productivity, interactions, pollination, water regulation) and structure (spatial organization of the species, number of species, population conditions).

According to SEMARNAT (2000), the vegetation of the forest areas including northern Mexico has been affected by several factors which may favor its degradation. The main factors are: clearing (94%), forest fires (2.1%), changing land-use authorization (1.3%), clandestine felling (1.0%), forest pests and diseases (0.5%) and other (1.0 %).

Based on the foregoing statements and as for the case of northeastern Mexico, disturbances can be classified into the following types:

Anthropogenic activities. Human activities increase the risk of degradation of vegetation by disturbance associated with intensive management done in soils with little or no knowledge at all of its dynamics, resulting in an accelerated deterioration of the ecosystem (Pritchett, 1986, González et al., 1990). The effect of anthropogenic activities in the scrub land in northeastern Mexico can be divided into those caused by the clearing of the vegetation for agriculture, livestock grazing and extraction of forest soil and wood (Maass, 1993).

Effect of agriculture on the structure of the scrub land. The main effect of agriculture on the structure of the scrub is when clearing is carried out stripping the vegetal cover completely to make way for agriculture. The loss of the scrub surface as a result of clearing for agriculture produce multiple negative effects on the environment such as habitat alteration, changes in biodiversity, presence of pests and diseases, alteration of nutrient cycles, changes in physical and chemical properties of the soil, contamination of surface and groundwater with nutrients and biocides, and soil loss (Finegan, 1993).

Effect of grazing on the structure of the scrubland. The effect of grazing on the structure of the shrub is mainly focused on the defoliation caused by livestock. The reduction of the leaf area produces a decline in the levels of photosynthesis by plants which limit the ability to produce organic carbon. On the other hand, reducing the transpiration surface due to grazing promotes the conservation of soil moisture longer and so extend the period of vegetative growth of plants. In places where water is available at the surface, areas free of canopy favor the increase of direct evaporation from the soil and can increase the levels of heat stress.

Figure 1. Left: farming activities in an area of submontane scrub, Right: Soil loss in an area of submontane scrub.

Figure 2. Installing contour lines to reduce soil loss; Right: Soil loss caused by agricultural land.

Figure 3. Left and Right. Clearing in submontane scrub areas.

Grazing intensity influences the horizontal distribution of biomass of shrubs. Species that are heavily browsed generally have a greater homogeneity in the horizontal distribution of biomass than those receiving lower grazing pressure, although for a longer period of time homogeneity of the first depends on complex processes of grass-herbivore coevolution (Finegan and Sabogal, 1998).

The absence of grazing in the early stages of succession favors herbaceous perennial erect growth habit, while prostrate or rosette and annuals tend to disappear. In overgrazed areas prostrate plants forms dominate and concentrate biomass near the soil surface.

FOREST PLANNING

Uncontrolled extraction of natural resources for forest harvesting reduces soil nutrients, productivity and biodiversity and fragment native vegetation. Management practices alter the productivity of the soil and can affect the soil surface with consequent effects on carbon sequestration and emissions of greenhouse gases (Gómez-Pompa et al., 1972 ; Gálvez, 1993).

Fire

Fire is a recurring natural phenomenon that regulates and promotes various processes of energy exchange of ecosystems. These events have occurred with varying intensity through time. But today the presence of fire awakens a strong concern among the population about the danger posed directly on human settlements, which have gradually invaded the wild, and indirectly, the effects on global climate by increasing carbon dioxide production as a product of combustion and even the destruction of large masses of vegetation that contribute to the collection and the production of oxygen (Muñoz-Robles, et al., 2005).

Erosion and Soil Fertility

Soil erosion remains a major problem affecting our country since each year eroded surfaces increases while forest surface is reduced considerably (SEMARNAT, 2001).

This is why the issue becomes important when native vegetation over a given area is partially or totally destroyed for whatever reason. It is when we need the whole circle of knowledge to be able to reverse the process that causes the site affectation.

In an interesting study, González and Rocha (2009) stress the importance of the family Fabaceae due to the existence of 8 species (which besides are multipurpose) that have a high reproductive plasticity and adaptability to changing conditions and provide plenty of organic matter, offer good coverage, recovery of soil, produce timber and have the ability to fix nitrogen. The most outstanding species are *Acacia constricta* and *Acacia rigidula* due to their physiological characteristics and the environmental services they provide to restore soil fertility. In the same condition is *Karwinskia humboldtiana* as a species for potential use in soil restoration. Moreover, in a study of the semi-arid vegetation in Valle del Mezquital, it is believed that mesquite, acacia and mimosa trees and cacti compose together the structural basis of these plant communities. They also foster biological interactions and have a significant influence on the development of the community composition and dynamics of the vegetation since these species act as nuclei for plant colonization and shelter for wildlife (Noss, 1987; Mackinnon, 1990; Machlis, 1993).

Low Temperatures

Synnott (1986) collected a review of the frosts that caused damage to the native vegetation in northeastern Mexico and highlighted the extent of damage of those frost events that occurred in 1925, 1940, 1951, 1962, 1963 and 1967. To this list we must add the frost of 1983, 1987 y1989 that also caused damage to vegetation in the region (Foroughbakhch and Heiseke, 1990).

These low temperatures occurring in winter periods in the Northeast of Mexico cause severe damage to native vegetation. The frosts that occasionally reach temperatures up to -12 ° C can cause damage to vegetation mainly on leaves and stems; the effect depends on the species and age of the tree which may lead to death. However, the speed of the response to this effect is variable depending on the species concerned.

In a study made by Carrillo (1990) about the deciduous thorny scrub it was showed that of those native species of trees and shrubs severe damaged by frost winter of 1989 more than 26% corresponded to *Cercidium macrum*, *Cordia boissieri*, *Pitecellobium ebano*, *Celtis pallida, Acacia rigidula, Acacia berlandieri* and *Condalia hookeri.* Tree and shrub species less affected by low temperatures were (<20%) *Dyospiros texana*, *Bernardia myricaefolia*, *Porliera angustifolia*, *Amiris texana*, *Forestiera angustifolia, Castela texana, Leucophylum frutescens, Bumelia celastrina* and *Acacia schaffneri*. However, the species that managed to recover faster (2% damage) in a period less than six months were *Porliera angustifolia*, *Amiris texana*, *Forestiera angustifolia, Castela texana, Leucophylum frutescens, Bumelia celastrina, Acacia schaffneri* and *Eysenhardtia polistachya*.

Rainfall Interception

The interception of rain water is part of the sub atmospheric water content as it is precipitation, evaporation and transpiration. The interception affects infiltration, runoff and soil evaporation. The rainfall hampered by vegetation is called intercepted precipitation (Belmonte and Romero, 1999).

The interception of rain water is another function of the bush. Total plant runoff represent 78.1% of total precipitation, which means that the water deposited on the soil of this community fulfill 18.9% (Navar et al., 1999 .) Yáñez-Díaz et al. (2010) determined that the interception of shrub vegetation is different in areas with *Eucalyptus*. They also found that in the canopy of *Ebenopsis ebano* the interception of rainfall is higher with a percentage of 23%. Canopies dominated by *Helietta parvifolia* showed an interception of 18% while for *Eucalyptus* sp the interception value was 12% and 10% in *Prosopis glandulosa*. The species *Acacia berlandieri, A. rigidula* and *Diospyros texana* have interception values of 18, 15 and 22% respectively (Cantú-Silva and González-Rodríguez, 2005).

Species with Potential Use for Scrubland Restoration

Many native trees and shrubs in Mexico can be potentially valuable for use in reforestation, ecological restoration and agroforestry development by tolerating various forms of management and cultivation in one or more systems of land use in warm-humid, temperate-cold, and arid-semiarid regions. Unfortunately, in practice the activities mentioned only propagates a very small fraction of the total biodiversity of trees and shrubs native to Mexico in forest nurseries. The reasons for not using a greater number of trees and shrubs of the approximately 9,000 existing species in the Mexican national flora are diverse. Some causes that have limited the spread and use of many native trees and shrubs in reforestation, ecological restoration and agroforestry development are due to ignorance of the biology and management of the diaspora that they produce (Niembro-Rocas, 2001).

Although recovery in most of the altered areas will not be possible to achieve up to how it was before, it is still possible to induce the development of protective vegetation which will help to preserve and increase soil fertility and diversity of plants and animals. A key resource to achieve this is by the use of native herbaceous and woody plant species with potential to grow in areas profoundly affected and, over time, allow the recovery of soil fertility, microclimate and hydrological cycle similar to the original and the restoration of at least part of the flora and fauna that still survive in some places. To this day, reforestation programs developed by state, military and federal government agencies have mainly employed exotic tree species known worldwide and some native species poorly known biologically which has prevented some success for the purposes mentioned. In contrast, several species of trees and shrubs of the genus *Caesalpinia, Enterolobium*, *Gliricidia, Guazuma,*and *Leucaena* living in Mexico have been used for reforestation and agroforestry restoration in other continents like Africa.

When reforesting with exotic species the problems of domestication and availability of propagules are solved, however, the results obtained require us to rethink the need to domesticate and learn to propagate native species for which it is necessary to make an inventory of species with the biological and ecological properties most suitable for every climate and environmental condition of the country.

To make a successful use of native species in development programs of agroforestry systems, ecological restoration and reforestation is essential to deepen our knowledge about the biology, ecology, propagation and management of the species available to enable domestication and develop efficient propagation techniques, and even get improvements on sexual selection, genetic engineering, cloning or some of its most valuable features. It is important to take into consideration the utility of native species to local people so this will result in better conservation of the restored areas.

Annex I is a list of important species in ecological restoration in northeastern Mexico presented according to their characteristics and/or specific application to the occurring disturbed ecosystem.

REFERENCES

Alba-Cuellar N. 2008. Manual Práctico de Reforestación. Primera edición. Grupo Latino Editores. 950 p.

Belmonte S.F. and D.A. Romero. 1999. Intercepción en algunas especies del matorral mediterráneo. Universidad de Murcia. 202 p.

Brown S. and A. Lugo. 1990. Tropical secondary forest. *Journal of tropical ecology* 6:1-31.

Cantú-Silva I. and H. González-Rodríguez. 2005. Pérdidas por intercepción de la lluvia en tres especies del matorral submontano. CIENCIA UANL VIII (001): 80-85.

Carrillo A. 1990. Efecto de algunos tratamientos silvícolas y factores abióticos en el manejo y regeneración del matorral. Thesis. Facultad de Ciencias Forestales, UANL. Linares, Nuevo León. 73 p.

Sendy T. 1987. The value of corridors (and design features of same) and small patches of hábitat. In: Nature conservation: the role of corridors. Saunders, Hobbs,Surrey and Sons. USA. 357-359.

FAO-UNESCO. 1980. Ecosistemas de los bosques tropicales húmedos: informe sobre el estado del conocimiento. Paris, Francia. 771 p.

Finegan B. 1993. Procesos dinámicos en bosques naturales tropicales. Curso de bases ecológicas para la producción sostenible. CATIE, Turrialba, Costa Rica. 25 p.

Finegan B. and C. Sabogal. 1988. El desarrollo de sistemas de producción sostenible en bosques tropicales húmedos de bajura. Un estudio de caso en Costa Rica. Parte 1. El Chasqui 17:3-24.

Foroughbakhch R. and D. Heiseke. 1990. Manejo Silvícola del Matorral: raleo, enriquecimiento y regeneración controlada. Reporte Científico No. 19. Facultad de Ciencias Forestales, Universidad Autónoma de Nuevo León, Mexico. 28 pp.

Gálvez J. 1993. Caracterización, diagnóstico y propuesta de manejo de los recursos naturales renovables en la zona del ejido municipal de Flores, Petén. Facultad de Agronomía, USAC. Guatemala. 255 p.

Gálvez J. 2002. La restauración ecológica: conceptos y aplicaciones. Instituto de Agricultura, Recursos Naturales y Ambiente. Guatemala. 23 p.

Gómez-Pompa A., C. Vázquez-Yáñez and S. Guevara. 1972. The tropical rain forest: a nonrenewable resource. *Science* 177: 762-765.

González, D. M. and D. L. Rocha. 2009. Evaluación de Especies Arbóreas y Arbustivas del Matorral Espinoso Tamaulipeco para Restauración Ecológica en el Noreste de México. Facultad de Ciencias Forestales, UANL.

González, K.V., G. Hernández and D. Hernández. 1990. Evaluación de la pérdida del suelo (1ª Aproximación) en una región de la serranía de Actopan, Hidalgo, México. In: Memorias del Primer Simposium Nacional "Degradación del Suelo"

Kondolf G.M. 1995. Five elements for effective evaluation of stream restoration: Restoration Ecology, vol. 3: 133–136.

Kondolf G.M. and E.R. Micheli. 1995. Evaluating stream restoration projects: Environmental Management, vol. 19: 1–15.

Lamprecht H. 1990. Silvicultura en los trópicos. Los ecosistemas forestales en los bosques tropicales y sus especies arbóreas. Posibilidades y métodos para un aprovechamiento sostenido. *GTZ*. 335 p.

Lieberg P. 1993. Strategies for population reintroduction; effects of genetic variability on population growth and size. *Conservation Biology* 7(1):194-199.

Maass M. 1993. México y Centro América: las peores tasas de degradación de suelos. En: Green Peace América Latina. No. 4:2.

Machlis, G. 1993. Áreas protegidas en un mundo cambiante: Los aspectos científicos. In: Parques y progreso. IUCN, BID. IV Congreso mundial de parques y áreas protegidas, Caracas, Venezuela. pp 37-53.

Mackinnon J. 1990. Manejo de áreas protegidas en los trópicos. IUCN-PNUMA. 314 p.

Muñoz Robles, C.A., E.J. Treviño Garza, J. Verástegui Chávez, J. Jiménez Pérez, and O. Aguirre Calderón. 2005. Desarrollo de un modelo especial para la evaluación del peligro de incendios forestales en la Sierra Madre Oriental de México. Boletín del Instituto de Geografía UNAM, 56: 101-117.

Navar J., F. Charles and E. Jurado. 1999. Spatial Variations of Interception Loss Components by Tamaulipan Thornscrub in Northeastern Mexico. *Forest Ecology and Management* 124 (2-3): 231-239.

Niembro-Rocas, 2001. Las Diásporas de los árboles y arbustos nativos de México: Posibilidades y limitaciones de uso en programas de reforestación y desarrollo agroforestal. *Madera y Bosques* 7(2):3-11.

Noss R. 1987. Corridors in real landscapes: a reply to Simberloff and Cox. *Conservation Biology* 1(2): 159-164.

Pritchett, W. L. 1986. Suelos forestales. José Hurtado Vega (Ed.). Limusa. México, D.F. pp. 36-44 and 87-102.

Richards, R.T., J.C. Chambers and C. Ross. 1988. Use of native plants on federal lands: Policy and practice. *J. Range Management*, 51: 625-632.

SEMARNAT (Secretaría del Medio Ambiente y Recursos Naturales). 2000. El sector forestal de México: Situación actual y perspectivas. Sistema Nacional de Información Forestal.

SEMARNAT (Secretaría del Medio Ambiente y Recursos Naturales). 2001.Mexico, segundo lugar a escala mundial en pérdida de suelo.

Society for Ecological Restoration International Science and Policy Working Group. 2004. The SER International Primer on Ecological Restoration. www.ser.org and Tucson: Society for Ecological Restoration International.

Sol-Sánchez, A., C.E. Zenteno-Ruiz, L.F. Zamora-Cornelio and E. Torres-Reyes. 2002. Modelo para la Restauración Ecológica de áreas alteradas. Kuxulkab' Revista de divulgación. 7(14):48-60.

Synnott, T., 1986. Efectos de las heladas de 1983 sobre las plantas de la región de Linares, Nuevo León. Reporte Científico No. 5. Facultad de Ciencias Forestales, Universidad Autónoma de Nuevo León. Mexico. 13 p.

IUCN, PNUMA, WWF. 1991. Cuidar la tierra: Estrategia para el futuro de la vida. Gland, Switzerland. 258 p.

Yáñez-Díaz M.L., I. Cantú-Silva, H. González-Rodríguez and J.I. Uvalle-Saucedo. 2010. Intercepción de la lluvia en tres especies de matorral submontano y una plantación de eucalipto en Linares, Nuevo León, México. VII Simposio Internacional sobre la Flora Silvestre en Zonas Aridas. 620-630.

ANNEX 1

I.- Species characteristics of communities of Pine Forest (900-1,000 m, trees 8-22 m tall) Oyamel and Abies (2,000 to 2 ,00 m, trees 15 to 25 m in height):

Pinus arizonica Engelm.
Pinus arizonica stormiae Martínez
Pinus ayacahuite brachyptera Shaw

Pinus cembroides Zucc.
Pinus culminicola Andersen et Beaman
Pinus greggii Engelm.
Pinus helepensis Mill.
Pinus hartwegii Lindl.
Pinus montezumae Lamb.
Pinus nesoni Shaw
Pinus pinceana Gordon
Pinus pseudostrobus Lindl.
Pinus pseudostrobus estevezi Martínez
Pinus rudis Endl.
Pinus teocote Dchl. et Cham.
Pinus teocote macrocarpa Show.
Pseudotsuga flahaulti Flous
Pseudotsuga macrolepis Flous
Pseudotsuga rehderi Flous
Juniperus flaccida Schltdl
J. deppeana Steud
Arbutus xalapensis Kunth.
Abies vejari (Mart.) Liu.

II.- Oak forest (550 a 900 m, tres 12-15 m tall)

Quercus affinis Schneid.
Q. almarensis C.H. Muell
Q. Brayi Small.
Q. canbyi Trel.
Q. chihuahuensis Trel.
Q. rysophylla H. et B.
Q. clivicola Trel. et Muller.
Q. cordifolia Trel.
Q. cuperata Trel. Et Muell.
Q. derrumbaderoensis G.H. Muell.
Q. diversifolia Neé
Q. edwarsae C.H. Muell.
Q. endlichiana Trel.
Q. emoryi Torr.
Q. intrincata Trel.
Q. flocculenta C.H. Mill.

Q. fusiformis Small
Q. gambelii Nutt.
Q. glaucocides Mart. et Gal.
Q. grevesii Sudw.
Q. greggii Trel.
Q. invaginata Trel.
Q. monterreyensis Trel. Et Muell.
Q. polymorpha Schl. Et Cham
Q. reticulate H. et B.
Q. rugulosa M. et G.
Q. rysophylla Weatherb.
Q. saltillensis Trel.
Q.standleyi Trel.
Q.tenuiloba C.H. Muell.
Q. vaseyana Buckl.

III.- Native species from desert zones

Dasylirion texanurn Scheele
Hechtia glomerata Zucc.
Berberis trifoliolata Moric.
Agave lechuguilla Torr.
Agave striata Engelm.
Agave bracteosa S. Watrs. Ex Engelm.
Echinocactus texensis Hoppfer
Echinocereus poselgeri Lem.
Echinocereus enneacanthus Engelm.
Echinocactus platyacanthus Link and Otto
Euphorbia antisyphilitica Zucc.
Ferocactus hamatacantus Britob and Rose
Ferocactus pilosus
Larrea tridentata (DC.) Coville
Mammillaria heyderi Muehlenpf.
Opuntia engelmanii Salm-Diek
Opuntia leptocaulis DC
Opuntia rnicrodasys (Lehm.) Pfeiff.
Schaefferia cuneifolia A. Gary
Sclerocactus scheeri Salm.Dick
Stenocereus griseus (Haw.)Buxb.

Thelocactus bicolor Briton and Rose
Thelocactus setispinusa (Engelm.) Anderson

IV.- Native species fron semi-arid zones

Acacia farnesiana (L) Willd
Acacia rigidula Benth.
Amyris texana (Buckl.) Willson
Berberis chochoco Schlecht
Bernardia myricaefolia (Scheele) Wats.
Caesalpinia mexicana A. Gray
Celtis pallida Torr.
Citharexylum berlandieri Robinson
Condalia hookeri M.C. Johnst.
Cordia boissieri DC.
Croton fruticulosus Engelm.
Diospyros palmeri Eastw.
Diospyros texana Scheele
Erhetia elliptica DC.
Eysenhardtia polystachya (Ortega) Sarg.
Forestiera angustifolia Torr.
Gymnosperma glutinosum
Helietta parvifolia (Gray) Benth.
Hibiscus cardiophyllus A. Gray
Karwinskia humboldtiana (Roem. et Schult) Zucc.
Lantana achyrantifolia Desf.
Lantana macropoda Torr.
Lantana velutina Mart. E Gal.
Melochia tomentosa L.
Mimosa malacophylla A. Gray
Neopringlea integrifolia (Hemsl.) S. Wats.
Opuntia leptocaulis DC.
Opuntia sp
Parthenium lozanianum Barthett
Pithecellobium flexicaule (Benth.) Coulter
Pithecellobium pallens (Benth.) Standl.
Porlieria angustifolia (Engelm.) Gray
Prosopis glandulosa Torr.
Salvia ballotaeflora Benth.

Schaefferia cuneifolia A. Gray
Zanthoxylum fagara (L.) Sarg.
Ziziphus obtusifolia (T.E. G.) Gray

V.- Native species from grasslands

Agropyron spicatum Jones
Agrostis stolonifera L.
Andropogon barbinodis (Trin.) Gould
Aristida barbata Fourn.
Aristida divaricata Eumb.et Bompl.
Arundo donax L,
Avena fatua L.
Bouteloua aristidoides H.B.K.
Bouteloua barbata Lag,
B. curtipendula (Michx.) Torr.
B. gracilis (H.B.K) Lag.
B. trifida Thurb.
Brachiaria meziana Hitch.
Brachypodium mexicanum R.et S.
Bromus purgans L.
Cenchrus brownii R.et S.
Chloris radiata (L.) Swartz
Cymbopogon oitratus (DC.) Stapf.
Digitaria adscendens (HBK)Henr.
Eragrostis curvula (Schrad.) Nees
Festuca arizonica Vasey
Hilaria belangeri (Steud.)Nash
Muhlenbergia repens (J. Prest.) Hitch.
Panicum hallii Vasey
Penisetum ciliare (L.) Link
Setaria corrugata (Elliot) Ex Schult
Sorghum halepense (L.) Pers.
Sporobolus airoides Torr.
Sporobolus virginicus (L.) Kunth
Tragus berteronianus Schult.
Trichloris pluriflora Foum
Tridens flavus (L.) Hitchc.

In: Native Species
Editors: L. Marin and D. Kovač

ISBN: 978-1-61470-613-7

Chapter 4

FLOWERING, FRUITING AND LEAF AND SEED VARIABILITY IN *BERBERIS BUXIFOLIA*, A NATIVE PATAGONIAN FRUIT SPECIES

Miriam E. Arena,[1*] Edgardo Giordani[2] and Silvia Radice[3]

[1]Laboratorio de Recursos Agronómicos, Centro Austral de Investigaciones Científicas (CADIC – CONICET), Bernardo Houssay, Ushuaia, Tierra del Fuego - Argentina

[2]Department of Plant, Soil and Environmental Science, University of Florence, Sesto Fiorentino (FI) Italy

[3]CEFYBO-CONICET-UBA, Paraguay, Buenos Aires, Argentina

ABSTRACT

Recent findings observed on *Berberis buxifolia* plants grown from different spontaneous populations on Tierra del Fuego island (Argentina) are reported, as related to: (I) the floral morphology and evolution of the floral phenological phases; (II) the fruiting efficiency; (III) the pattern of

---

* Author for Correspondence: Miriam E. Arena, Laboratorio de Recursos Agronómicos. Centro Austral de Investigaciones Científicas (CADIC – CONICET) Bernardo Houssay 200 (9410) Ushuaia, Tierra del Fuego – Argentina. Tel.: +54 2901 422310. Fax: +54 2901 430644. E-mail address: arena1@infovia.com.ar; arena@cadic-conicet.gob.ar.

fruit growth and evolution of some of the chemical attributes; and (IV) the variability found in the leaves and seeds. Flowers are solitary, axillary, bisexual, actinomorphic and hypogynous. Petals present the nectaries characteristics of the genus. Fruit is a black-blue berry (8-10 mm diameter) with an average of 3-18, moon-like shaped seeds. Bloom occurs at different times, and its beginning and duration is an accession characteristic, depending also on ecological factors. Nevertheless, seven floral phenological phases were recognized. Phenophases of blooming were measured during the spring of 2010. The length of blooming of the *B. buxifolia* accessions studied was 15 days between start of blooming and petal fall. Fruit number per shoot varied with the orientation and during the fruiting, being higher in the north (4.3) and west (4.0) orientations than in those from the east (2.0) and south (1.3), attaining maxima of 8.8 and 7.8 fruits per shoot respectively at day 38. Fruit efficiency then began to decrease during the ripening, attaining values of 20 to 25% for those shoots from the north and west orientations, and values of 13 to 15% for the shoots from the east and south. Fruit growth showed a typical double sigmoid curve, and it was related to compositional changes. By day 126 from full flowering phase, soluble solids (30.6 °Brix) and anthocyanin content (752.7 mg/100 g fruit fresh weight) were near their maximum values, while the total titratable acidity was at a minimum value (2.2 %). The present investigation confirmed the previous description of leaves; nevertheless a considerable variability has been observed in populations of *B. buxifolia*. Leaf size ranged from 0.04 to 2.08 cm^2 (average 0.5 cm^2); similar results were found for leaf shape (major axis/minor axis of leaf blade ranged from 1.26 to 6.25); furthermore, leaves color analysed with the RGB scale varied among populations and plants. In *B. buxifolia* seed number per fruit may range from 1 to 15, with an average 7 seeds/berry; results indicated that clonal selection may be applied for this trait. Seed area ranges from 5 to 16 mm^2, with an average of 9 mm^2; seed are lunular, mostly elongated and very elongated in shape.

Introduction

In spite of the well known importance of wild flora as source of food and medicinal substances, only a few species have been evaluated for their agronomic and medicinal potential (Iriondo, 2001). Hence, more study on the wild flora diversity and on the agronomic and medicinal potential of these plant species is still needed (Arena and Vater, 2005). Areas with much indigenous flora offer non-domesticated plants (Monge et al., 2000), such as *Berberis* L. genus in Patagonia, which is represented by 16 species of native shrubs (Orsi, 1984), with a wide distribution from Neuquén Province to Tierra del Fuego island (Job, 1942; Orsi, 1984). At present, commercial orchards of some *Berberis* species are being planned, as it represents an attractive crop, first because its black-blue fruits can be consumed fresh and processed in marmalades and jams (Orsi, 1984), in non-alcoholic beverages, and in ice creams; and secondly, because it is an important source of alkaloids i.e. berberines, and phenolic compounds such as anthocyanins, which offers medicinal and tinctorial applications (Pomilio, 1973; Shaffer, 1985; Fajardo Morales et al., 1986; Fajardo Morales, 1987). *Berberis buxifolia* Lam. (locally called "calafate") is an evergreen, spiny and erect shrub up to 4 m high, often growing in coastal scrub, *Nothofagus* forest margins and clearings, moister areas in grass steppes, and along streams and rivers (Moore, 1983). *Berberis buxifolia* is considered a non timber forest product (Tacón Clavaín, 2004), that can be propagated through seeds (Arena and Martínez Pastur, 1994), rhizomes (Arena and Martínez Pastur, 1995; Arena et al., 1998) and *in vitro* culture (Arena et al., 2000). Shoot growth and fruit production were studied in a natural population of this species (Arena et al., 2003).

Knowledge about blooming and pollination is indispensable to maximizing the cropping potential of fruits (Nyéki and Solstész, 1996). In addition, the complete study of the structure and morphology of the flower such as the development, viability and functional capacities of the sexual organs, accompanied by a description of the flowering, pollination and fertilization processes, are essential to initiate an improvement program. Currently, few studies in these areas with regard to *Berberis buxifolia* are available in the literature (Orsi, 1982). Nevertheless, in other species of the *Berberidaceae* family, detailed investigations are reported. Floral morphology and development of the gametophytes was described by Sastri (1969) in *Berberis umbellata*. Some aspects of *Berberidaceae* pollen were investigated by Kosenko (1980), while information about stamen irritability was published by Grieve in 1931, even if previously described by Linnaeus (Grieve, 1931).

The study of the evolution of fruit size and weight, in correlation with components such as soluble solids, and those giving acidity and pigments, could be of value in defining the optimal time for harvesting *B. buxifolia* fruits according to their future use, while contributing to the knowledge of their nutraceutical properties at a particular stage of maturity (Gonzalez-San José et al., 1990).

Furthermore, interpopulation variability found in the leaves and seeds, is helpful for studying the species gene pool, hence it is relevant when defining a selection strategy aiming at the improvement of the species.

Currently, floral biology, fruit development and the assessment of morphological variability are the main gaps in the description of this species. Recent findings observed on plants grown from different populations on Tierra del Fuego island (Argentina) are reported in the present work and they are related to (I) floral morphology and evolution of the floral phenological phases, (II) fruiting efficiency, (III) pattern of fruit growth and evolution of some of the chemical attributes such as soluble solids, acidity, and anthocyanin content, and (IV) the variability found in leaves and seeds.

Figure 1. Floral morphology of *Berberis buxifolia*. **A**, flower organs, anther (a), petal (p), gynoecium (g), sepal (s), stigma (st); **B**, detail of the petal nectarie; **C**, non dehiscent anther; **D**, dehiscent anther; **E**, young pistil taken from flower in phase E; **F**, detail of stigma with pollen grains; **G**, ovules in basal placentation. Bars = 1 mm.

FLOWERING, FRUITING AND VARIABILITY OF LEAVES AND SEEDS

(I) Floral Morphology and Evolution of the Floral Phenological Phases

Flowers are solitary, axillary, bisexual, actinomorphic and hypogynous. Calyx, with two cycles of 3 sepals, is free and petaline yellow colored (Figure 1A). Corolla, with 6 free petals, is yellowish orange; each petal brings 2 nectaries in basal position (Figure 1B). Androecium, with 6 stamens opposite to the petals, is bithecous (Figure 1C), dehiscent along longitudinal lateral valves, opening from base upwards and the connective with cleared apex (Figure 1D). Pollen grains are tricolpate and numerous. Gynoecium is monocarpellar (Figure 1E), with a superior ovary, unilocular with numerous ovules, anatropous and with basal placentation (Figure 1G). Stigma is moist with abundant secretion (Figure 1F), being receptive before the beginning of anthesis and lasts two thirds of the time of the whole anthesis phase. Style is very short or absent (Figure 1E). Fruit is a black-blue berry 8-10 mm diameter (Figure 2) with an average of 3-18, moon-like shaped seeds (Figure 10).

Figure 2. Ripe *Berberis buxifolia* fruits. Bars = 1 cm.

Figure 3. Phenological phases of flower buds of *Berberis buxifolia*. **A**, flowers buds (arrows) in phase D; **B**, phase E previous to anthesis; **C**, flower buds in different phases, detail of F_2 phase (arrow head); **D**, anthesis F_1; **E**, flowers at petal fall (phase G); **F**, young fruit (H). Bars **A, C** = 10mm. **B, D, E, F** = 1 mm.

Bloom occurs at different times, and its beginning and duration is an accession characteristic, and depend also on ecological factors. Seven floral phenological phases (D, E, F_1; F_2, F_3, G, H) have been recognized during the blooming period according to a modification of the Fleckinger scale (Fleckinger, 1955). The floral button enlarges very rapidly and fifteen days after its apparition, anthesis occurs. Green flower buttons appear between the basal rosettes of foliage leaves and quickly the pedicel enlarges and the flower button increases its diameter (phase D, Figure 3A). When the pedicel elongation finishes, flower buttons change to yellow color (phase E, Figure 3B). In this phase, stamens show still non dehiscent anthers, close to the stigma. Anthesis starts when the sepals and petals move from the center towards the outside leaving exposed the pistil (phase F_1, Figure 3D). Afterwards, sepals move to a position of 90° with respect to the floral axis whereas the petals surround partly the reproductive organs. Stamen move as well as petals (phase F_2, Figure 3C). Petal senescence is a characteristic of the end of anthesis (phase F_3, Figure 3E). The length of the anthesis phase is closely related to the climatic conditions; in particular, in 2010 the blooming period was recorded as having a duration of 21 days (Figure 4). Petal fall (phase G) is a very short phase (only 2-4 days), since the fruits are suddenly exposed (phase H, Figure 3F). Fifteen days after the beginning of the bud-break, all the flowers were in the petal fall phase.

Phenophases of blooming were measured during the spring of 2010 (Figure 4). Start of bloom was considered when 10 % of flowers were in F_1 phase (about October 27). Full bloom, considered as 50 % of flowers in anthesis phase, was recorded on October 29; the end of blooming was registered on November 9, when 20 % of flowers were in petal fall (Figure 4).

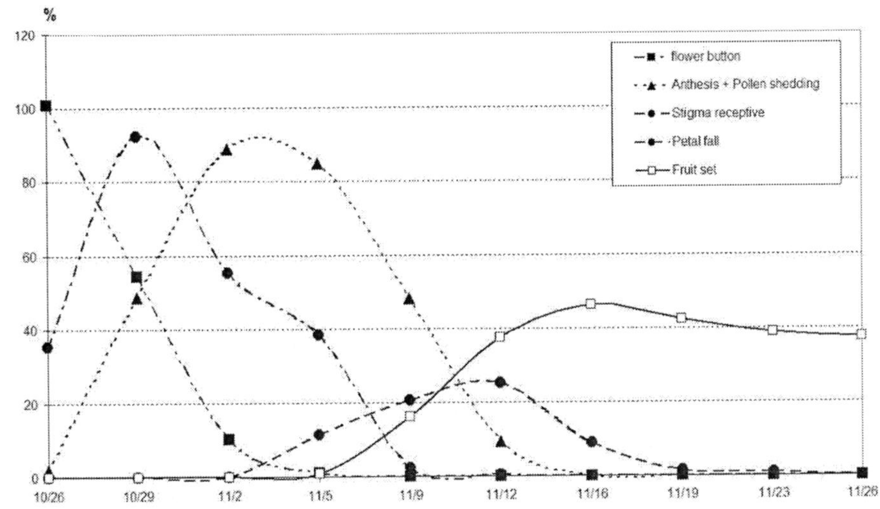

Figure 4. Blooming phenogram of *Berberis buxifolia* recordered in the spring of 2010. Values are expressed in percentage with respect to the total number of flower buds sprouted.

Table 1. Mean values of ANOVA for the analysis of mixed bud number per shoot *(MBN)*, mixed bud percentage per shoot *(MBP)*, fruit number per shoot *(FN)* and fruit percentage per shoot *(FP)* in *Berberis buxifolia* considering the orientation and duration in days of the flowering and fruiting periods as main factors *(n=11)*

Main effects	MBN	MBP	FN	FP
Orientation				
East	4.41b	24.73b	2.01b	13.22b
North	7.36a	33.94a	4.28a	16.93a
South	3.30b	25.02b	1.26b	9.98b
West	6.85a	32.05a	3.97a	17.96a
Days				
1	16.57a	66.47a	0.00e	0.00f
6	16.31a	67.95a	0.00e	0.00f
13	14.86ab	70.44a	0.04e	0.18f
17	14.13ab	69.36a	0.31e	1.38f
20	13.15abc	67.79a	0.72e	3.43ef
24	11.54bc	64.00a	1.56def	8.41def
27	9.61c	58.63ab	2.56bcdef	13.11bcde
31	5.40d	47.64b	4.70ab	23.37ab
34	1.79de	24.94c	5.45a	26.83a
38	0.56e	9.39d	5.52a	27.46a
42	0.22e	3.19d	5.31a	26.78a
50	0.00e	0.00d	5.11ab	25.52a
63	0.00e	0.00d	4.68ab	23.52ab
77	0.00e	0.00d	4.56ab	22.82ab
90	0.00e	0.00d	4.43abc	22.29ab
104	0.00e	0.00d	3.71abcd	19.39abc
118	0.00e	0.00d	3.40abcde	17.17abcd
132	0.00e	0.00d	1.84cdef	9.65cdef
146	0.00e	0.00d	0.83ef	4.61ef

Significance of the main effects O: Orientation; D: Days. *MBN:* O= 0.0000, D= 0.0000, Interaction 0.0000; MBP: O= 0.0000, D= 0.0000, Interaction 0.1492; FN: O= 0.0000, D= 0.0000, Interaction 0.0495; *FP:* O= 0.0000, D= 0.0000, Interaction 0.9359. Values followed by different letters in each column and for each factor are significantly different with Tukey Multiple Range test at $p<0.05$.

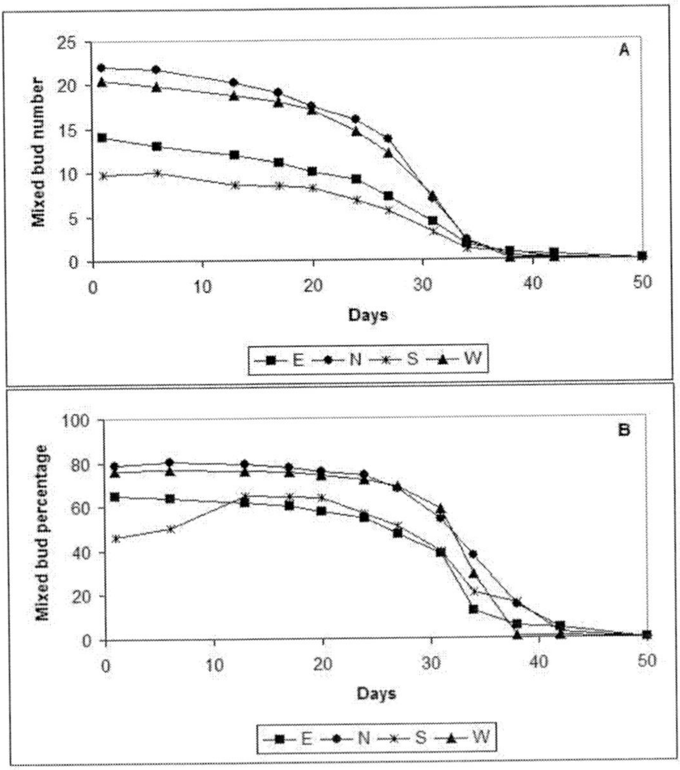

Figure 5. Mixed bud evolution in *Berberis buxifolia*, depending on shoot orientation and duration in days of the blooming and fruiting periods. **A.** Mixed bud number per shoot and **B.** Mixed bud percentage with respect to the total bud number per shoot. E (East), N (North), S (South), W (West).

(II) Fruiting Efficiency

The fruiting shoot characteristics and fruiting efficiency were studied during the flowering and fruiting periods, from the opening of the mixed and vegetative buds until the end of the fruit ripening. One-year-old shoots (n=11), with 19.5 ± 3.5 cm length, ANOVA $p \geq 0.05$, and 17.8 ± 3.5 nodes, ANOVA $p \geq 0.05$, were chosen taking into account the plant and shoot orientation (north, south, west or east). The number of mixed buds per shoot, containing both vegetative and reproductive organs, varied with the shoot orientation and during the flowering and fruiting periods, being higher in the north (7.4) and west (6.8) orientations than in those from the east (4.4) and south (3.3) (Table

1, Figure 5A). The same behavior was observed when the mixed buds were expressed as relative number (number of mixed buds with respect to the total buds of the shoot) (Table 1, Figure 5B).

The fruits appeared first in the north, east and west orientations, and later in the south orientation (Figure 6). Fruit number per shoot varied with the orientation and during the study period, being higher in the north (4.3) and west (4.0) orientations than in those from the east (2.0) and south (1.3) (Table 1), attaining maxima of 8.8 and 7.8 fruits per shoot respectively at day 38 (Figure 6A).

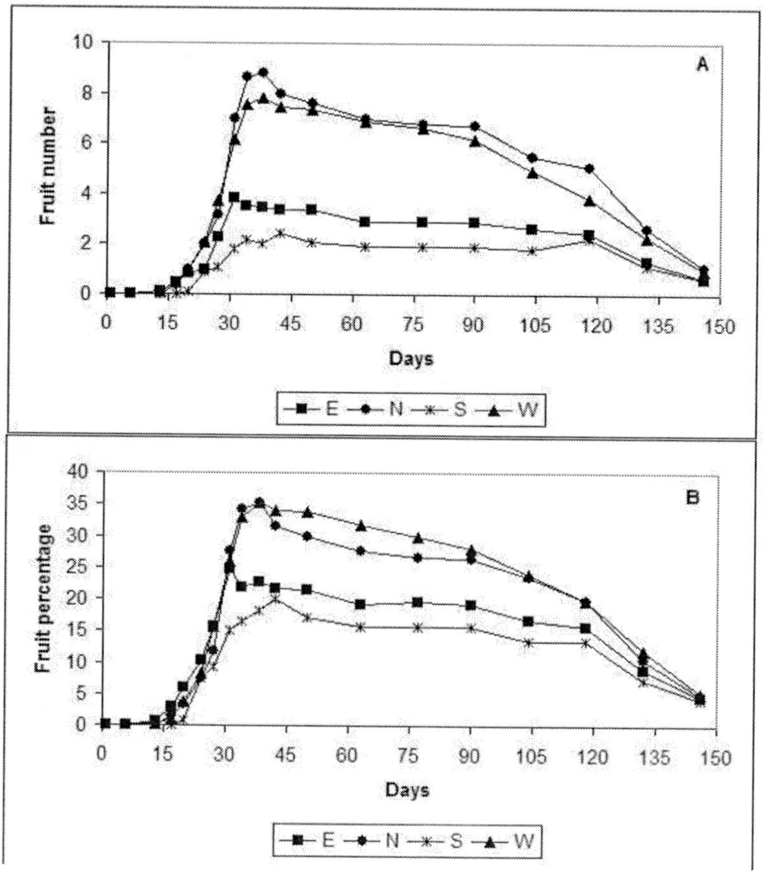

Figure 6. Fruit production in *Berberis buxifolia*, depending on shoot orientation and duration in days of the blooming and fruiting periods. **A.** Fruit number per shoot and **B.** Fruit percentage with respect to mixed bud number per shoot. E (East), N (North), S (South), W (West).

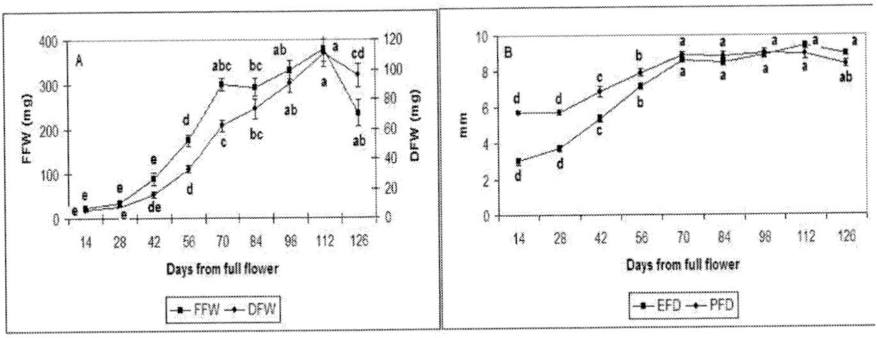

Figure 7. Measurements recorded in the *Berberis buxifolia* fruits during the fruiting period. **A**. Fresh (*FFW*) and dry (*DFW*) fruit weight. **B.** Equatorial (*EFD*) and polar (*PFD*) fruit diameters. Values followed by different letters in each variable are significantly different with Tukey Multiple Range test at $p<0.05$. *(n=20)*.

The same behavior was observed when the fruit number was expressed as relative number (number of fruits with respect to the number of mixed buds) (Table 1, Figure 6B). Two weeks after the petal fall phase (day 38), the reproductive efficiency (number of fruits with respect to the number of mixed buds per shoot) was near 35% in those shoots from the north and west orientations, being significantly higher than the shoots from the east and south (22.7 and 18.2% respectively). Fruit efficiency then began to decrease during the ripening period, attaining values of 20 to 25% for those shoots from the north and west orientations, and values of 13 to 15% for the shoots from the east and south orientations; these decreases could be attributed to a natural fruit drop. During the ripening phase, even if not an intrinsic factor, it is worth noting that a significant decrease occurred in the fruit number due to attack by birds, with 10 to 30% of the ripened fruit damaged. The higher number of mixed buds and fruits obtained in the north and west orientations compared to the east and west could be due to a longer exposure to the sun, and consequently light intensity that is too high, as well as to a less exposure to the predominant winds from the south-south west (Arena et al., 2003).

(III) Pattern of Fruit Growth and Evolution of some of the Chemical Attributes

Sun-exposed fruits were collected from plants growing naturally in an area located near Ushuaia city, 54° 48′SL, 68° 19′WL (Tierra del Fuego, Argentina), from November 2007 (14 days after full flowering phase) to

March 2008 (126 days after full flowering phase). Fresh fruit weight significantly varied during the fruiting period (p=0.000), particularly during the first growth phase (Figure 7A). Fruit growth (fresh weight basis), over the course of time, from day 14 after full flowering phase until day 112, shows a typical double sigmoid curve of growth, with the highest fruit growth rate around days 42 and 56 after full flowering phase (10.5 and 16.1% per day, respectively), reaching a lag phase by day 70 after full flowering phase (300.2 mg), which was extended until day 84. During the third phase of rapid increase until 112 days after full flowering phase (time when the fruits also ripen), the highest biomass was obtained (379.8 mg), and then a decrease in fresh biomass (averaging 2.7% per day) occurred toward the end of the growing season. The dry fruit weight significantly changed during the fruiting period (p=0.000) (Figure 7A). The time course evolution on the basis of dry fruit weight also showed a double sigmoid curve with a less pronounced plateau, and increased slightly to reach an average biomass of 111.1 mg per fruit by day 112 after full flower phase. The typical double sigmoid curve of growth shown by *B. buxifolia* was also observed in stone fruits (Coombe, 1976; Gil-Albert Velarde, 2006), as well as in fruits of *Vitis* (Coombe and McCarthy, 2000) and *Ribes* (Toldam-Andersen and Hansen, 1997). The dry/fresh fruit weight ratio reached a maximum mean value of 41.2% by day 126. The equatorial and polar fruit diameter changed significantly during the fruiting period (p= 0.000 and p= 0.000, respectively), increasing from day 14 after full flowering phase to their maximum mean values of 9.4 mm (112 days after full flowering phase) and 9.0 mm (98 days after full flowering phase) for equatorial and polar fruit diameter respectively (Figure 7B). These variables then decreased toward the end of the growing season, following the same pattern as the fruit weight did. Fruit firmness significantly changed during the fruiting period, being at its maximum at the beginning of the fruiting period, to then decrease towards the end of this period with mean values of 0.15 and 0.19 kgf by days 112 and 126 after full flowering phase, respectively.

Soluble solids significantly increased during the ripening phase, to reach a maximum (30.6 °Brix) by day 126 after full flowering phase (Table 2); these values are comparable to those obtained by Arribillaga García (2001). Total titratable acidity significantly decreased during the ripening period (Table 2) with a maximum (6.3%) by day 70 after full flowering phase, to then decrease towards the end of ripening. As expected, the soluble solids/total titratable acidity ratio significantly increased during the ripening phase, with a maximum (13.9) by day 126 after full flowering phase (Table 2). Anthocyanin content significantly increased during the ripening phase, reaching a maximum by day 119 after full flowering phase (762.7 mg/100 g fresh fruit weight). Anthocyanin production began by day 84 after full flowering phase and then markedly increased during the ripening period, when fruit biomass and dry fruit weight, expressed as percentage, were at their maximum values. It then increased until the end of the fruiting period even when fresh fruit mass decreased. Thus, in *B. buxifolia* biosynthesis of anthocyanin compounds starts after a pronounced lag phase from fruit setting.

The precise conditions that initiate anthocyanin synthesis are not well established, but it has been reasonably argued that sugar accumulation provides the substrate needed for anthocyanin synthesis, i.e. the sugars accumulated beyond the immediate need for fruit tissue functional integrity, often favor the synthesis of secondary metabolites such as anthocyanins, and various flavorants (Jackson, 2008). The anthocyanin content found in *B. buxifolia* fruits at maturity (762.7 mg/100 g fresh weight) was comparable to the obtained in earlier growing seasons (Arena and Curvetto, 2008), and higher than the obtained in *B. buxifolia* in Chile (500 mg/100 g fresh weight) (Arribillaga García, 2001).

Table 2. Soluble solids (*SS*, ºBrix), total titratable acidity (*TTA*, %), soluble solids/total titratable acidity relation (*SS/TTA*), and anthocyanin content (*A*, mg/ 100 g FFW) in *Berberis buxifolia* fruits during the days after full flowering phase (*AFFP*)

Days AFFP	SS (ºBrix)	TTA (%)	SS/TTA	A (mg/100 g FFW)
70	10.96i	6.31a	1.72g	1.74e
77	11.43h	4.72b	2.41fg	19.88e
84	13.36g	4.03c	3.32fg	35.78e
91	14.33f	3.66c	3.90ef	102.63d
98	14.86e	2.79d	5.33de	165.71c
105	19.43d	2.83d	6.89cd	254.64b
112	19.93c	2.45d	8.10bc	316.66b
119	23.86b	2.47d	9.68b	762.70a
126	30.56a	2.19d	13.95a	752.71a
p	*0.0000*	*0.0000*	*0.0000*	*0.0000*

ANOVA (p) = probability at p=0.05. Values followed by different letters in each column are significantly different with Tukey Multiple Range test at p<0.05. *(n=6)*.

The anthocyanin content in this study was also higher than that of other reddish-purple berries, such as blackcurrants (350 mg/100 g FW), raspberries (55-60 mg/100 g FW), strawberries (40 mg/100 g FW) (Burrows and Moore, 2002; Lister et al., 2002), and blueberries, equal to 1.2 mg/g FW (Zheng and Wang, 2003) and 149 to 176 mg/100 g FW (Divo de Sesar et al., 2008).

When analyzing the changes in fruit growth together with the evolution of the chemical properties during the studied fruiting period, it can be concluded that parameters such as fruit biomass and size attain the highest values earlier (98 to 112 days from full flowering phase) than chemical properties (119 to 126 days from full flowering); this fact must be taken into account when defining optimum time of harvest depending on intended fruit utilization.

(IV) Variability Found in the Leaves and Seeds Obtained from Plants Grown from Different Populations on Tierra del Fuego Island (Argentina)

The leaves of the species belonging to the genus *Berberis* L. are botanically classified as simple, fasciculate, from membranous to strongly coriaceous and from entire to multispinulose; *B. buxifolia* is reported to have elliptic, entire, leathery, persistent, leaves (Orsi, 1984). More specifically, leaves of this species are up to 10-fasciculated, alternate, from sexile to subsexile, glabrous, coriaceous, with up to 1.5 mm pubescent petiole and 10-20 x 5-9 mm leaf blade, decurrent, elliptic or obovate, obtuse, mucronated, attenuated in the base, glossy and papillous in the adaxial blade and dull in the abaxial side; swollen and entire margin (Orsi, 1984). Despite this botanical description, confirmed by recent field surveys, considerable variability has been observed in spontaneous populations of *B. buxifolia* on Tierra del Fuego island. Broad patterns of leaf surface characteristics have been previously related to ecophysiological performance (Brewer and Nuñez, 2007) and to divaricating habit (McQueen, 2000) in *Berberis* from Argentine Patagonia.

Leaf size ranged from 0.04 to 2.08 cm^2 (average 0.5 ± 0.25); the lowest value of perimeter was 1.25 cm, the highest 7.67 cm, with an average of 3.57 ± 1.0 cm (Figure 8). Studying a wide number of specimens belonging to different populations and growing in different environments, the analysed factor "population/location" and "plant" resulted statistically significant for these two variables (ANOVA; $p<0.01$). Similar results were observed for the parameters calculated to describe leaf shape, such as major axis/minor axis of leaf blade (1/elongation), that ranged from 1.26 to 6.25 (average 2.56 ± 0.6), which corresponds to obovate-elliptic leaf shape (Figure 8).

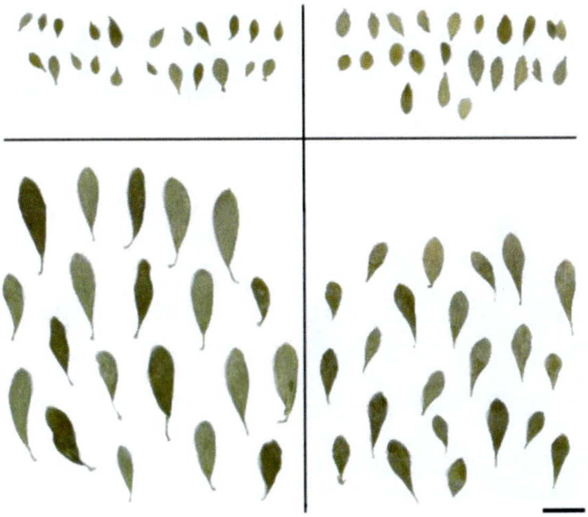

Figure 8. Variability in leaf size and shape of *Berberis buxifolia* populations on Tierra del Fuego island. Bar = 1 cm.

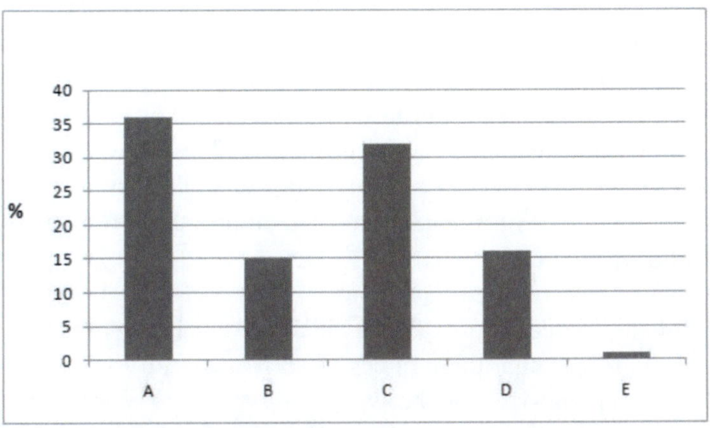

Figure 9. Distribution (%) of plants of *Berberis buxifolia* based on leaf blade color and on electronic colorimeter. A (R =63 ± 2; G = 71 ± 1.2; B = 48 ± 2.6); B (R =56 ± 2.6; G = 63 ± 2.7; B = 44 ± 1.2); C (R =68 ± 1; G = 77 ± 2; B = 52 ± 4); D = (R =86 ± 7; G = 93 ± 6; B = 57 ± 5); E = (R =124; G = 120; B = 63.

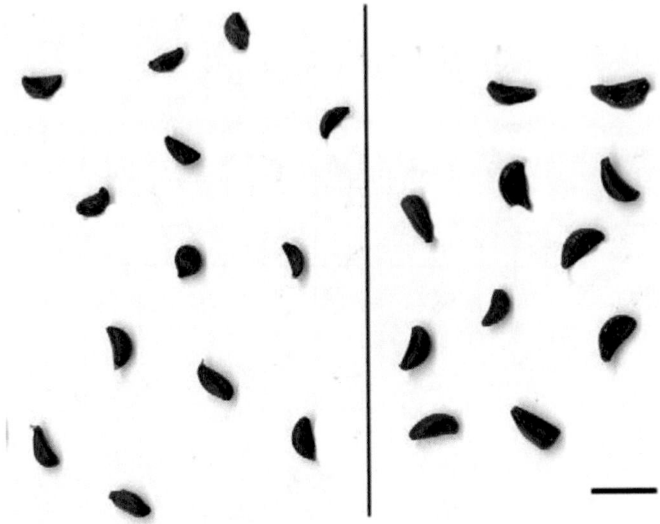

Figure 10. Variability in seed size and shape of *Berberis buxifolia* populations on Tierra del Fuego island. Bar = 1 cm.

The variation of adaxial leaf blade color is quite wide. Based on the RGB color scale (Colorimeter PCE – RGB Model), the average of R, G and B were 67.41 ± 20, 74.87 ± 20 and 50.27 ± 15, respectively. Figure 9 reports the distribution of plants by leaf color, observed in different populations of *B. buxifolia* on Tierra del Fuego island, after counting the occurrences in different color groups obtained by Cluster Analysis (Euclidean Distance, Group Average; $p>0.001$) employing R, G and B as variables.

New leaf emergence occurs in early spring, while leaves at end their life fall in early autumn, with great differences among the different locations on Tierra del Fuego island.

In genus *Berberis*, seeds are lunular, crustacean, dark brown or bluish black; from the scarce information available from literature on *B. buxifolia* seeds, it is evidenced that the number of seeds per fruit ranges from 7 to 11, being 5 x 2 mm in size and brown in color (Orsi, 1984) (Figure 10). In addition to the reproductive function of seeds, their presence, number and morphology are relevant pomological characteristics which may affect edibility and suitability to processing calafates. There are neither reports nor direct observations about the existence of seedless berries of *B. buxifolia*. On the contrary, the number of seeds per fruit ranges from 1 up to 15, with a general average of 7 ± 2.7 seeds/berry. Taking into account a set of 95 different native plants belonging to 25 populations of *B. buxifolia* spread in different locations of Tierra del Fuego island, the seed number per fruit was statistically affected both by population/location and plant factors (ANOVA; $p<0.01$). Aside from environmental and ecological factors (such as the presence of pollinators), which should in any case be better studied, these observations indicate that clonal selection for low seed number per fruit is possible within *B. buxifolia*. The average number of seeds/fruit in each plant varied from 2.3 ± 1.1 to 11 ± 2.

A notable variability of seed dimension is also present in this species. Seed area ranges from 5 to 16 mm^2, with an average of 9 ± 1.7 mm^2; seed length (major axis), varies from 0.32 to 6.9 mm (average of 0.44 ± 0.16 mm); lower mean values (0.23 ± 0.03 mm) were observed for the width of the longest cross section, ranging from 0.05 to 0.43 mm. The average seed weight of *B. buxifolia* is 12.7 ± 2.5 mg, with seeds ranging from 6.9 to 18.6 mg. In the previously cited calafate population, seed area and weight are influenced both by location and plant factors (ANOVA; $p<0.01$), hence selection of plants with fruits having small seeds seems to be achievable.

Seed are mostly elongated to very elongated in shape; the width/length ratio ranges from 1.16 (almost round) to 4.8 (very elongated), with a mean value of 1.9 ± 0.3. Other calculated parameters such as roundness [$(4 \pi \times \text{area})/\text{perimeter}^2$)] and compactness, computed as sqrt $(4\text{xarea}/\pi)$/major axis length, confirm these observations.

CONCLUSION

Knowledge about blooming and pollination is indispensable to maximize the cropping potential of fruits. Moreover, complete study of the structure and morphology of the flower, such as the development, viability and functional capacities of the sexual organs, accompanied by the description of the flowering, pollination and fertilization processes, are essential to initiate an improvement program. Differences in floral morphology were not observed between the accessions studied, meanwhile blooming time depended on each studied plant. The study of fruit growth evolution in correlation with components such as soluble solids, and those giving acidity and pigments, can be considered as good markers of the fruiting phases and ripening, contributing to the knowledge of their nutraceutical properties at a particular stage of maturity. Finally, the interpopulation variability found in the leaves and seeds is helpful for studying the species gene pool, hence it is relevant when defining a selection strategy aiming at the improvement of the species.

REFERENCES

Arena, M.E. and Curvetto, N. (2008). *Berberis buxifolia* fruiting: Kinetic growth behavior and evolution of chemical properties during the fruiting period and different growing seasons. *Scientia Horticulturae 118*, 120-127.

Arena, M.E. and Martínez Pastur, G. (1994). Seed propagation of *Berberis buxifolia* Lam. *Phyton 56(XII)*, 59-63.

Arena, M.E. and Martínez Pastur, G. (1995). Propagación de frutales menores nativos de los bosques Andino-Patagónicos: El Calafate. *Presencia X (37),* 5-7.

Arena, M.E., Vater, G. and Peri, P. (1998). Propagación por rizomas de los *Berberis* de la Patagonia Austral. Proceedings IX Congreso Latinoamericano de Horticultura; XLIX Congreso Agronómico de Chile: 96, Santiago, Chile.

Arena, M.E., Martínez Pastur, G. and Vater, G. (2000). *In vitro* propagation of *Berberis buxifolia* Lam. *Biocell 24 (1)*, 73-80.

Arena, M.E., Vater, G. and Peri, P. (2003). Fruit production of *Berberis buxifolia* Lam. in Tierra del Fuego. *HortScience 38(2),* 200-202.

Arena, M.E. and Vater, G. (2005). Native and introduced species of small fruits in Austral Patagonia, Tierra del Fuego. In: R. Dris (ed), *Fruits: Growth, Nutrition, and Quality* (WFL Publisher, Helsinki, Finlandia). ISBN: 952-99555-0-2. 220 pp. http:///www.isfae.org/book/php

Arribillaga García, D. (2001). Domesticación del calafate (*Berberis buxifolia* Lam.) para fines agroindustriales. INIA, Chile. 137 p.

Brewer, C.A. and Nuñez, C.I. (2007). Patterns of leaf wettability along an extreme moisture gradient in Western Patagonia, Argentina. *International Journal of Plant Science 168 (5),* 555-562.

Burrows, C. and Moore, P.P. (2002). Genotype x Environment effects on raspberry fruit quality. *Acta Horticulturae 585,* 467-473.

Coombe, B.G. (1976). The development of fleshy fruits. *Annual Review of Plant Physiology 27,* 507-528.

Coombe, B.G. and McCarthy, M. (2000). Dynamics of grape berry growth and physiology of ripening. *Australian Journal of Grape and Wine Research 6 (2),* 131-135.

Divo de Sesar, M., Pepe, F., Massara, D., Kaen, R., Lopez, V., Stella, A. and Vilella, F. (2008). Variación en el contenido de polifenoles y antocianas en los frutos de los géneros *Vaccinium* y *Rubus* asociada a diversos factores condicionantes. In: M. Divo de Sesar et al (eds.).Avances en cultivos frutales no tradicionales (153-165). Buenos Aires, Argentina.

Fajardo Morales, V., Podestá, F. and Urzúa, A. (1986). Reseña de los alcaloides encontrados en el género *Berberis* de Chile. *Rev. Latinoamericana de Química 16,* 141-156.

Fajardo Morales, V. (1987). Estudio químico de las especies chilenas del género *Berberis*. *Rev. Latinoamericana de Química 18,* 46-50.

Fleckinger, J. (1955). Phenologie et arboriculture frutière. *Bon Jardinier 1,* 362-372.

Gil-Albert Velarde, F. (2006). Tratado de arboricultura frutal. Vol 1: Morfología y fisiología del árbol frutal. Ed Mundi Prensa, España, 104 pp.

Gonzalez-San Jose, M.L., Barron, L.J.R. and Diez, C. (1990). Evolution of anthocyanins during maturation of Tempranillo grape cultivar (*Vitis vinifera*) using polynomial regression models. *Journal of the Science of Food and Agriculture 51,* 337-344.

Grieve, M. (1931). A Modern Herbal V. 1, p. 82.

Iriondo, J.M. (2001). Conservación de germoplasma de especies raras y amenazadas. *Revista de Investigación Agraria, Serie: Producción y Protección Vegetales 16 (1),* 5-24.

Jackson, R.S. (2008). Wine science: principles and applications. Elsevier Ed. 751 p.

Job, M.M. (1942). Los *Berberis* de la región del Nahuel-Huapi. *Rev del Museo de la Plata (Sección Botánica) 5,* 21-72.

Kosenko, V.N. (1980). Comparative palyno morphological study of the family Berberidaceae. 2. Morphology of the pollen grains of the genera *Gymnospermium, Leontice, Caulophyllum, Bongardia, Epimedium, Vancouveria, Achlys* and *Jeffersonia. Botanicheskii zhurnal (Moscow and Leningrad) 65,* 1412-1423.

Lister, C.E, Wilson, P.E., Sutton, K.H. and Morrison, S.C. (2002). Understanding the health benefits of blackcurrants. *Acta Horiculturae 585,* 443-449.

Mc Queen, D.R. (2000). Divaricating shrubs in Patagonia and New Zealand. *New Zealand Journal of Ecology 24 (1),* 69-80.

Monge, A., Chorghade, M., Erhardt, P.W., Genellin, C.R., Koga, N., Lindberg, P., Perun, T.J., Topliss, J.G., Trivedi, B.K., Wermuth, C.G. (2000).

Medicinal chemistry in the development of societies: Biodiversity and natural products. *European Journal of Medicinal Chemistry 35,* 1121-1125.

Moore, D.M. (1983). Flora of Tierra del Fuego. Oswestry, England. Anthony Nelson and Missouri Botanical Garden. 396 p.

Nyéki, J. and Soltész, M. (1996). Floral biology of temperate zone fruit trees and small fruits. Akadémiai Kiadó, Budapest.

Orsi, M.C. (1984). Berberidaceae. *In*: Correa, M. N. *ed.* Flora Patagónica. Secc. 4a. Tomo VIII. INTA. Buenos Aires, Argentina. Pp 325-348.

Pomilio, A.B. (1973). Anthocyanins in fruits of *Berberis buxifolia. Phytochemistry 12,* 218-220.

Sastri, RLN. (1969). Floral morphology, embryology, and relationships of the Berberidaceae. **Australian Journal of Botany** *17(1),* 69-79.

Shaffer, J.E. (1985). Inotropic and chronotropic activity of berberine on isolated Guinea pig Atria. *Journal of Cardiovascular Pharmacology 7,* 307- 315.

Tacón Clavaín, A. (2004). Manual de Productos Forestales No Madereros. Valdivia, Chile. CIPMA.

Toldam-Andersen, T. and Hansen, P. (1997). Growth and development in black currants (*Ribes nigrum*). III. Seasonal changes in sugars, organic acids, chlorophyll and anthocyanins and their possible metabolic background. *Journal of Horticultural Science 72 (1),* 155-169.

Zheng, W. and Wang, S.Y. (2003). Oxygen radical absorbing capacity of phenolics in blueberries, cranberries, chokeberries, and lingonberries. *Journal of Agriciculture and Food Chemistry 51,* 502-509.

In: Native Species
Editors: L. Marin and D. Kovač
ISBN: 978-1-61470-613-7

Chapter 5

A CONTRIBUTION TO THE NATIVE EARTHWORM SPECIES OF ASSAM STATE IN NORTHEAST REGION OF INDIA

Shweta Yadav
Vermiculture Research Station, Dept. of Zoology
D.S. College (Dr BRA University), Aligarh, India

ABSTRACT

Assam is a high earthworm diversity region and constitutes 26.2% of the total known Indian species; among them 22% are native of India. The earthworms have innate capability to convert various industrial wastes into nutrient rich vermicompost with higher agronomic value besides enhancing soil fertility. A survey of earthworm bioresources with particular reference to native species was, therefore, undertaken during 2010 in Guwahati, Goalpara and Bangaigaon districts of Assam, hitherto, uninvestigated in this direction. The survey provided a total of six species of earthworms, namely, *Lampitomauritii*Mauritii, *Perionyxexcavatus* Perrier, *Metaphireanomola*Michaelsen, *Drawidanepalensis*Michaelsen, *Ocnerodrilusoccidentalis*Eisen and *Metaphireposthuma*Vallant; of these later two species are exotics.

The study indicates that *Lampitomauritii*Mauritii and *Metaphireposthuma*Vallant are the predominant species for the area investigated. Need for such investigations with a view to target native species of earthworms and conserve their biodiversity in underexplored or unexplored areas of India and abroad has been re-stressed.

Keywords: Biodiversity, Earthworm, Assam, Northeast region (NER) of India, Vermicompost

INTRODUCTION

The diversity of earthworms may be influenced by altering the physico-chemical properties of soil replacement of forest and natural savannas with agro-ecosystems (Chaudhuri et al., 2008). Studies in Mexico, Peru and India reveal that agro-ecosystem earthworm communities (as compared with those in the undisturbed ecosystem) have lower species richness, lower numbers of native species and lower number of ecological groups (Chaudhuri,and Bhattacharjee, 2005). However, the role of earthworms in enhancing primary production depends on the synlocalization (in the same place) and the synchronization (at the same time) of their activities with the period and the sphere of active root growth and nutrient demands.

Earthworms account for the highest biomass among tropical soil macro fauna (Fragosoand Lavelle, 1992). The role of earthworms in the process of decomposition, building and maintenance of soil structure has been well documented for soils of temperate region (Edwards and Bohlen 1996; Lee 1985). In tropics studies on diversity, ecology and role of earthworms have been carried out in savanna (Lavelle, 1974), pasture (Dash andPatra, 1977), rain forests (Fragosoand Lavelle, 1992) and agricultural lands (Sathianarayanan and Khan, 2006). However, there is paucity of information on earthworm communities in subtropical forests (Fragoso et al., 1999; Gonzalez *et al.*, 2007).

Northeast region of India, comprises eight states, namely, Arunachal Pradesh, Assam, Manipur, Meghalaya, Mizoram, Nagaland, Sikkim and Tripura and lies between 22°05□ N and 29°30□ Nlatitude and 87° 55□ E and 97° 24□ Elongitude. The region is characterized by diverse agro-climatic and geographical conditions. Valleys are rich in organic matter. Because of continuous removal of topsoil, organic matter status is poor to medium on the steep slopes. It occupies 7.7% of India's geographical area supporting 50% of the flora and fauna, of which 31.58% is endemic (Yumnam, 2008). It is a transitional zone between the Indian, Indo-Burman, Malaysian and Indo-Chinese regions. The rich biodiversity of NER has a significant role in the maintenance of the ecosystem. However, intensive mining and continuous degradation of natural resources (soil, water, vegetation) are disturbing their fauna and flora (Chaudhuri and Bhattacharjee,2005;Chaudhuri et al.,2003). This region is, therefore, facing shrinkage of green cover everywhere, and experiencing an impact on its ecological system, leading to habitat loss and habitat fragmentation and ultimately resulted in biodiversity loss.

Indian Earthworm Taxonomic Diversity

Approximately 3700 species of earthworms are known in the world (Reynolds, 1994) but the number of estimated species may be as high as 8000 (Fragoso et al., 1997). In India, 418 species and subspecies belonging to 69 genera have been described so far (Julka and Paliwal, 2005). This number is expected to rise to about 800 with extensive surveys of large unexplored area of 'biodiversity hotspots of the Western Ghats and Eastern Himalaya' (Julka, 1988). High earthworm diversity in India is primarily due to its geographical location with a wide latitudinal range (between 8.4^0 and 37.6^0N), complex topography, varied climate (ranging from temperate to arctic in the Himalaya to tropical in the peninsular India) and past geological history that is linked to ancient super continent of Gondwanaland from which it separated in the late Jurassic and drifted to collide with the Asian mainland in the Eocene.

Indian Earthworm Resources

Julka (1988) demonstrated occurrence of nine families with 53 genera and more than 400 earthworm species of India particularly in Himalayan, Indo-Gangetic and Deccan peninsula. The Deccan peninsula is rich in earthworm fauna and harbours many epigeic and anecic species such as *Dichogasterbolaui,Drawidawillsi*, *Perionyxexcavatus*, *Perionyxsansibaricus*, *Raniella* sp., *Lampitomauritii* and other species which have potentiality for use in vermitechnology (Dash and Senapati, 1985; Dash, 1999). The north-east and Himalayan regions are also rich in many endemic and some exotic species of earthworms (Bhadauria and Ramakrishanan, 1989, 1991, Bhadauria et al., 2000, Sinha et al., 2003). The exotics are *Octochaetonatyrtaeum*, *Amynthascorticis*, *Bimostosparvus* and the endemic species are *Octochaetonabeatrix*, *Eutyphoeusfestiveus*, *E. nanianus*, *E. walton*iand *Drawida* species. Distinct taxonomic groups of earthworms have evolved on every continent except Antarctica. But through human transport, many exotic species have been imported from many other regions especially from Europe, Africa and America and some groups have been distributed worldwide (Jamieson, 1978; Reynolds and Cook, 1976). Earlier studies by Dash and Patra (1977), Senapati and Dash (1981), Mishra and Dash (1984), Senapati (1993) have shown that in pastures and mixed wood forest soils in Orissa, the earthworm species composition was not similar although *Lampitomauritii* was found in all sites (pastures, croplands and forestsites).

A review of literature on earthworm biodiversity in northeast India, however, indicates that after the important works on taxonomy and distributional survey by Stephenson (1923) and Gates (1972), very few studies are available for this region. These include Julka and Halder (1975), Julka (1976), Julka (1981, 1988), Julka et al. (2005), Chaudhuri and Bhattacharjee (2005), Bhadauria and Ramakrishnan (1989, 1991), Bhadhauria, Mishra and Ramakrishnan (1988). However, there appears no work available on the earthworm bioresources of Guwahati, Goalpara and Bangaigaon districts of Assam. Therefore, present studies on earthworm biodiversity survey were undertaken with a view to update existing information and contribute material for the preparation of earthworm fauna for Assam state in general and the study area in particular. The study presents systematic enumeration of six species of earthworms collected and identified from diverse habitats of the area investigated.

Rationale of the Study

Of recent, native earthworm species have received renewed attention of scientist in India and abroad because of their innate capability to convert various industrial wastes for *e.g.* sugar mill waste, timber waste, paper waste etc. into nutrient rich vermicompost with higher agronomic value, besides enhancing soil fertility. Based on this rationale, an exhaustive survey of earthworm bioresources with particular reference to native species was conducted in hilly, urban and waste land areas of Guwahati, Goalpara and Bangaigaon districts of Assam state in northeast region of India during 2010, hitherto, uninvestigated in this direction. The survey provided some 95 specimens of earthworms comprising six species collected and identified from diverse habitats (rural and hilly forest area, cultivated fields, waste lands) of the study area (Table 1 and 2).

Table 1. Survey details of earthworm bioresources in the study area

SI. No.	Voucher Specimen no.	Site of Collection	Habitat
1.	A/01	Guwahati University, Guwahati	Ungrazed grassland
2.	A/02	Institute for Information and Technology, Guwahati	Ungrazed grassland
3.	A/03	Kamakya Temple, Guwahati	Grazed grassland
4.	A/04	Nursery, Guwahati	Grassland with rich organic matter
5.	A/05	River bank of Brahamputra near M.G. road, Guwahati	Ungrazed grassland
6.	A/06	Raj Bahawan Road, Guwahati	Grazed grassland
7.	B/01	The Energy Research Institute (TERI), Guwahati	Ungrazed Grassland
8.	B/02	Range Forest Office, Indira Nagar Road, Guwahati	Forest
9.	B/03	Vijaypur Garden, Vijapur Road, Guwahati	Ungrazed Grassland
10.	B/04	Forest Area, Vijaypur Road, Guwahati	Forest
11.	B/05	Forest Area, Baristha Road, Guwahati	Forest
12.	B/06	Arts College Garden, Guwahati	Ungrazed Grassland
13.	C/01	Bharamputra river bank, Chalantapur, Bongaigaon Road, Goalpara	Ungrazed Grassland
14.	C/02	Jogighopa, Bongaigaon Road, Goalpara	Ungrazed Grassland
15.	C/03	Forest, Lakhimpur Road, Goalpara	Forest
16.	C/04	Dalgona, Malia Road Goalpara	Cultivated land
17.	C/05	Kharmuja, Dakurbhita Road Goalpara	Ungrazed Grassland
18.	D/01	Manas National Park, Bhutan Hill Road, Goalpara	Forest
19.	D/02	River bank, Manas, Bhutan Hill road, Goalpara	Ungrazed Grassland
20.	D/03	Royal Manas Park, Central Area, Bhutan Hill Road, Goalpara	Forest Hill Area
21.	D/04	DhumaniBarpeta Road, Goalpara	Cultivated land
22.	D/05	Fathemabad, Bongaigaon Road, Goalpara	Forest
23.	E/01	Chaprakata, Bongaigaon Road, Bongaigaon	Ungrazed Grassland

Table 1. (Continued)

SI. No.	Voucher Specimen no.	Site of Collection	Habitat
24.	E/02	Chapaguri, Dhaligaeen Road, Bongaigaon	Cultivated land
25.	E/03	Tulongia Market, Abhayapuri Road, Bongaigaon	Ungrazed Grassland
26.	E/04	Bidyapur, Basugen road, Bongaigaon	Ungrazed Grassland
27.	E/05	Bongaigaon road, River bank, Bongaigaon	Ungrazed Grassland

Table 2. Date-wiseSurvey Record of EarthwormsCollection from the Study Area

SI. No.	Collection Date(s)	Name of district of Assam state, India	Location	Voucher specimen no(s).	Habitat
01	29.07.2010-30.07.2010	Guwahati	$26^0 11$□ 0N $91^0 44$□0 E and 55m above sea level	A/01; A/02; A/03; A/04; A/05; A/06; B/01; B/02; B/03; B/04; B/05; B/06	Grazed and ungrazed grassland, land with rich organic matter, Forest
02	31.07.2010-01.08.2010	Goalpara	$26^0 17$□ 0N $91^0 62$□ 0E and 35m above sea level	C/01; C/02; C/03; C/04; C/05; D/01; D/02; D/03; D/04; D/05	Grazed and ungrazed grassland, cultivated land, Forest
03	02.08.2010-	Bongaigaon	$26^0 47$□ 0N $90^0 57$□ 0E and 54m above sea level	E/01; E/02; E/03; E/04; E/05	Cultivated land, ungrazed and grazed grassland

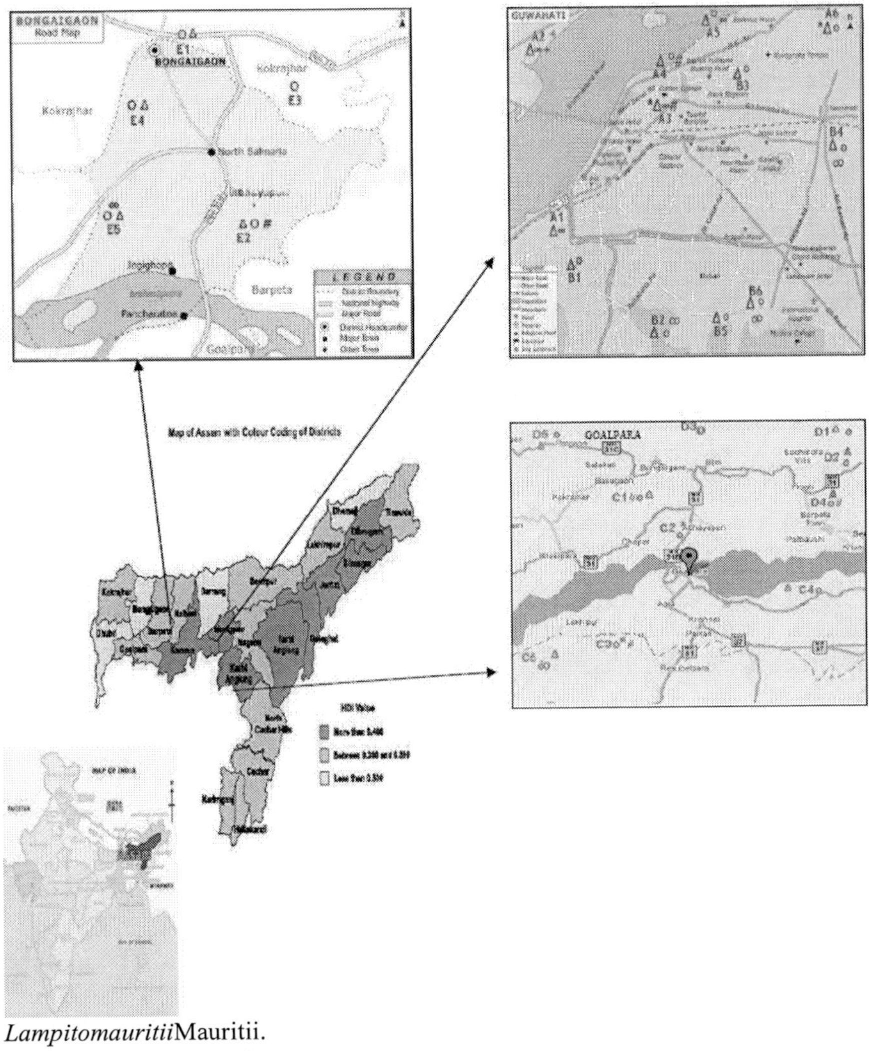

Δ *Lampitomauritii*Mauritii.
O*Metaphireposthuma*Vallant.
* *Metaphireanomalus*Michaelsen.
# *Perionyxexcavatus* Perrier.
∞ *Drawidanepalensis*Michaelsen.
\+ *Ocnerodrilusoccidrituls*Eisen.

Figure 1. Study Area of northeast region in India: showing distribution of different species of earthworms.

The Study Area

The study area (Guwahati, Goalpara and Bangaigaon districts) of Assam is situated between 26°11□0□ □N latitude and 9°144□ 0□ □ E longitude at altitude of 55 m above sea level (Figure 1). Total forest area of the state is 28,747 km^2, out of which 359 km^2 which bodies and covers 32% of the total geographical area (Patiri and Borah, 2007). The field visits included both urban and rural waste lands, forests and cultivated fields. Mean daily temperature ranges from 20° C in January to 32° C in July and precipitation averages 1600 mm annually.

Materials and Methods

Field work was carried out in the study area (Fig 1) during July and August, 2010. The methodology adopted for earthworms collection was based on Julka (1988). Earthworms and soil samples for taxonomic studies were collected by digging and hand sorting method. Samples were collected from diverse ecological niches viz. dense forests, ungrazed and grazed grasslands, cultivated lands, land with rich organic matter and river banks. Collected worms were washed in fresh water and stored in test tube in the field. Ethyl alcohol was gradually added to the test tube containing earthworms and then transferred to the dish containing solution of 5% formalin for fixation. This is kept for 6-8 hrs and later preserved in 70% ethyl alcohol or 5% formalin. All specimens collected were serially numbered and field data recorded in the field diary each day. All specimens collected from study area were examinedat the laboratory of Vermiculture Research Station (VRS), D.S. College, Aligarh, and were provisionally identified with the help of available literature and later confirmed at the Zoological Survey of India, Kolkata. Voucher specimens reported in the present work are deposited in the Museum of VRS, for future reference and study.

Experimental Design and Statistical Analysis

On each sampling site earthworms were hand-sorted from three soil monoliths (50x50x20 cm) placed at the apices of a 10 m side triangle. Specimens were preserved in 5% formalin, provisionally identified and counted. Relative density was calculated following Cousin's method (1991).

RESULTS

Adverting shortly to the presentation of data, all earthworm species collected and identified from the study area are arranged family-wise in alphabetical order. Each entry gives the information in sequence: earthworm's scientific name, origin, locality and collection number, date of collection, relative density and general habitat.

1. Family Megascolecidae

Its distributional range extends between warm-temperate Asia and Australia. Native species of *Lampitomauritii* have been recorded from almost all collection sites. Endemic Megascolecids in this region belongs to genus *Perionyx.* Species explosion seems to have occurred in the NER, may be due to considerable and regular rainfall and high organic matter in the soil(Chaudhuri and Bhattacharjee ,2005).

LampitoMauritiiMauritii

Origin: Native

Locality and collection (nos): *Guwahati:* A/01 Guwahati University, Guwahati; A/02 Institute for Information and Technology, Guwahati; A/03 Kamakya Temple, Guwahati; A/04 Nursery, Guwahati; A/05 River bank of Brahamputra near M.G. road, Guwahati; A/06 Raj Bahawan Road, Guwahati; B/01 The Energy Research Institute(TERI), Guwahati; B/02 Range Forest Office, Indira Nagar Road, Guwahati; B/03 Vijaypur Garden, Vijaypur Road, Guwahati; B/04 Forest Area, Vijaypur Road, Guwahati; B/05 Forest Area, Baristha, Road Guwahati ; B/06 Arts College Garden, Guwahati; *Goalpara:* C/01 Bharamputra river bank, Chalantapur, Bongaigaon Road, Goalpara; C/04 Dalgona, Malia Road, Goalpara; C/05 Kharmuja, Dakurbhita Road, Goalpara; D/01 Manas National Park, Bhutan Hill Road, Goalpara; D/02 River Bank, Manas, Bhutan Hill Road, Goalpara; D/04 DhumaniBarpeta Road, Goalpara; D/05 Fathemabad, Bongaigaon Road, Goalpara; *Bongaigaon:* E/01 Chaprakata, Bongaigaon Road, Bongaigaon; E/02 Chapaguri, Dhaligaeen Road, Bongaigaon; E/04 Bidyapur, Basugen Road, Bongaigaon; E/05 Bongaigaonroad,river bank, Bongaigaon.

Date(s) of Collection: 29.07.2010 – 02.08.2010

Relative Density: 65-80%

General habitat: Cultivated land, grassland (grazed, ungrazed), hill area, under stones, leaf litter, river bank

MetaphirePosthuma Vaillant

Origin: Exotic

*Locality and collection (nos):Guwahati:*A/04 Nursery, Guwahati; A/05 River bank of Brahamputra near M.G. road, Guwahati; A/06 Raj Bahawan Road, Guwahati; B/01 The Energy Research Institute (TERI), Guhawati; B/02 Range Forest Office, Indira Nagar Road, Guwahati; B/03 Vijaypur Garden, Vijaypur Road, Guwahati; B/04 Forest Area, Vijaypur Road, Guwahati; B/05 Forest Area, Baristha Road, Guhawati; B/06 Arts College Garden, Guwahati; *Goalpara:* C/01 Bharamputra river bank, Chalantapur, Bongaigaon Road, Goalpara; C/02 Jogighopa, BangaigaonRoad,Goalpara; C/03,Forest,Lakhimpur Road,Goalpara; C/04 Dalgona, Malia Road, Goalpara; D/01 Manas National Park, Bhutan Hill Road, Goalpara; D/02 River Bank, Manas Bhutan Hill Road, Goalpara; D/03 Royal ManasPark,CentralArea,Bhutan Hill road,Goalpara; D/04 DhumaniBarpeta Road, Goalpara; D/05 FathemabadBongaigaon Road, Goalpara; *Bongaigaon:* E/01 ChaprakataBongaigaon Road, Bongaigaon; E/02 ChapaguriDhaligaeen Road, Bongaigaon; E/03 TulongiaMarket,Abhayapuri Road, Bongaigaon; E/04 Bidyapur, Basugen Road, Bongaigaon; E/05 Bongaigaon road river bank, Bongaigaon.

Date(s) of collection: 29.07.2010 – 02.08.2010

Relative Density: 40-80%

General habitat: Cultivated land, grassland (grazed, ungrazed), river bank, leaf litter.

MetaphireAnomalaMichaelsen

Origin: Native

Locality and Collection (nos):Guwahati:A/01Guwahati University, Guwahati.

Date(s) of collection: 29.07.2010

Relative Density: 50-60%

*General habitat:*Cultivated land

PerionyxExcavatusMichaelsen

Origin: Native

*Locality and collection (nos):Guwahati:Guwahati:*A/03 Kamakya Temple, Guwahati; A/04 Nursery, Guwahati; *Goalpara:*C/01

Bharamputrariver bank, Chalantapur, Bongaigaon Road, Goalpara; C/03,Forest,Lakhimpur Road,Goalpara; D/04 DhumaniBarpeta Road, Goalpara;*Bangaigaon:*E/02 Chapaguri, Dhaligaeen Road, Bongaigaon.

*Date(s) of collection:*29.07.2010 – 02.08.2010

Relative Density: 55-60%

General habitat: Hill area, river bank, grassland (ungrazed, grazed), cultivated land.

2. Family Moniligastridae

This family comprises primitive earthworms of East and South Asia. A few species are hydrophilous and some are common in arable soils but most are confined to primary forests. Of the Indian genera *Desmogaster*, *Drawida* and *Moniligaster* are the largest genera in terms of number of species. Its natural distribution extends from the Peninsula to the Eastern Himalaya.

DrawidaNepalensisMichaelsen

Origin: Native

Locality and collection (nos):Guwahati: A/01 Guwahati University, Guwahati; A/02 Institute for Information and Technology, Guwahati; A/03 Kamakya Temple, Guwahati;

A/05 Brahamputra river bank near M.G. road, Guwahati; B/02 Range Forest Office, Indira Nagar Road, Guwahati; B/04 Forest Area, Vijaypur Road, Guwahati; B/06 Arts College Garden, Guwahati; *Goalpara:*C/05 Kharmuja, Dakurbhita Road; Goalpara*Bangaigaon:*E/05 Bongaigaon road river bank. Bongaigaon.

Date(s) of collection: 29.07.2010

Relative Density: 45-65%

General habitat: Cultivated land, grassland (ungrazed).

OcnerodrilusOccidentalisEisen

Origin: Exotic

Locality and collection (nos): Guwahati: A/02 Institute for Information and Technology, Guwahati.

Date(s) of collection: 29.07.2010

Relative Density: 45%

General habitat: Grassland (ungrazed)

Table 3. Earthworm Resources of the Study Area

SI No.	Earthworms	Guwahati													Goalpara											Bangaigon		
1.	Megascelecidae-Lampitomauritii	+	+	+	+	+	+	+	+	+	+	+	+	+	-	-	+	+	+	+	-	+	+	+	+	-	+	+
	Metaphire-posthuma	-	-	-	+	+	+	+	+	+	+	+	+	+	+	+	+	+	+	+	+	+	+	+	+	+	+	+
	Metaphirea-nomola	+	-	-	-	-	-	-	-	-	-	-	-	-	-	-	-	-	-	-	-	-	-	-	-	-	-	-
	Perionyxex-cavatus	-	-	+	+	-	-	-	-	-	-	-	-	+	-	+	-	-	-	-	-	+	-	-	+	-	-	-
2.	Moniligastridae Drawidanepalensis	+	+	+	-	+	-	-	+	-	+	-	+	-	-	-	-	+	+	-	-	-	-	-	-	-	-	+
	Ocnerodilus-occidentalis	-	+	-	-	-	-	-	-	-	-	-	-	-	-	-	-	-	-	-	-	-	-	-	-	-	-	-

+ present, - absent.

DISCUSSION

The role of earthworms (EWs) in soil fertility is known since 1881,when Darwin (1809–1882) published his book entitled "*The Formation of Vegetable Mould Through the Action of Worms with Observations on Their Habits*". Since then, several studies have been made to highlight the soil organism's contribution to the sustainable function of all ecosystems (Aune, 2009). Adding to this the EWs have received renewed attention of scientists in India and abroad because of their innate capability to convert various industrial wastes into nutrient-rich vermicompost with higher agronomic value, besides enhancing soil fertility. Therefore, search for native species of earthworms is on the top priority in the present day researches and some recent contributions in this direction are those of Verma et al. (2010) and Verma and Shweta (2011). However, there is an urgent need to study earthworm bioresources in different unexplored geographical areas of the country and elsewhere. This is important since urbanization and development of such areas particularly in North-East India is leading to enormous habitat loss of earthworms because of mining, deforestation, monocropping etc. Hence the present studies assume greater importance and calls for suitable measures for their conservation and protection in such areas.

Of the 95 field samples collected, the present study has brought to light presence of six species of earthworms belonging to six genera and two families from the study area. Of these four species, namely, *Lampitomauritii*Mauritii; *Perionyxexcavatus* Perrier; *Metaphireanomola*Michaclsen; *Drawidanepalensis*Michaclsen are native and other two *Oenerodrilusoccidentalis*Eisen and *Metaphireposthuma*Vallantareexotic in origin. *Lampitomauritii and Metaphireposthuma* are the predominant species with distribution in all collection sites. The structure of earthworm communities is determined by a suite of ecological factors: temperature, available organic matter, moisture, pH etc. Native earthworm species collected and identified from the study area have shown narrow edaphic and climatic tolerance, and hence, are stenoecious (Table 3). Low density of earthworms in the study area is probably related to the acidic nature of soil in Assam because earthworms, in general, are neutrophilic in nature (Lee, 1985).

In recent years, however, vermicomposting has become very popular all over the world and has gone modification keeping the availability of desired earthworm species and many other conditions. Most of the present day studies deal with three earthworm species viz.

Eudriluseugeniae,Eiseniafoetida,Perionyxexcavatus, for vermicomposting. The changes in C/N ratio, C/P ratio, concentration of NPK, phosphatase activity in worm casts, microbial interactions, and some other parameters have been taken as criteria for judging the efficiency of earthworms in the vermicomposting process. Reproductive fitness of the worm, growth, mortality factors based on feeding pattern have also received attention of research workers. It would be, therefore, be worthwhile to undertake detailed studies on such earthworms' species inhabitant to northeast by subjecting them to vermicomposting of organic portion of urban waste, agricultural waste, animal dung, water hyacinth, organic based industrial wastes from feed processing plants, sugar mill, paper mill, kitchen waste etc. and for field inoculation, as vermifeed and related purposes. As such, screening of potentiality of Indian earthworm species for vermitechnology purpose should be considered a thrust area for research.

Need for Protection and Conservation of Earthworm Biodiversity in North-east India

Earthworms (EWs) are a major component of soil fauna communities in most natural ecosystems and comprise a large proportion of macro fauna biomass. These play a key role in soil organic matter (SOM) transformation and nutrient dynamics at different spatial and temporal scales through perturbation and the production of biogenic structures for the improvement of soil fertility and land productivity (Brussaard et at., 1997; Levelle and Spain, 2001). With shrinkage of green cover everywhere, the northeast region is also experiencing an impact of its ecological system, thereby adversely affecting the earthworms' biodiversity. This calls for immediate concern of scientists and remedial steps for protection and conservation of EWs biodiversity in the region. In this direction following suggestions need attention: (i) There is immediate need to put a ban on over exploitation of forests, ill managed road construction, mining etc. which lead to habitat loss and habitat fragmentation that ultimately results in the loss of earthworm population; (ii) An immediate check on deforestation and restriction on developmental activities will also help preventing natural calamities such as landslides, floods and forest fires; (iii) Because of the deforestation and the resultant loss of soil especially in the hill areas, rivers and streams, the deep pools that are the favored habitats of many species of EWs, are rapidly becoming shallow and choked with salt, leading to a habitat loss and thereby reducing their population; (iv) The region is known for its age old institutional mechanisms on cultural and social values for biodiversity conservation, namely, *sacred groves or forests, sacred hill tops*. Conservation of such abode of local deities, representing remnants of ancient forests will help substantially protecting EWs biodiversity in ecosystems; (v) There is a need to check the process of biodiversity conversion through initiating awareness programmes among public to save existing biodiversity of earthworms. In such events, information on EWs life histories, field population variations ecological strategies and short and long term effects on soil properties in relation to crop production should be the priority; (vi) Future research must focus on describing the role of native species of EWs and their interaction with or replacement by exotic species in soil function and agro-ecosystems productivity.

ACKNOWLEDGMENT

The authors are thankful to the Department of Biotechnology [grant no. BT/PR9273/AGR/21/229/2007], Ministry of Science and Technology, Government of India, New Delhi, for financial assistance for the present study.

REFERENCES

Aune, J.B., 2009. Organic and conventional agriculture compared impacts on food production and environment. In: 4th world congress on Conservation Agriculture, New Delhi, India, pp 243.

Bhadauria, T., Ramakrishnan, P.S., Srivastava, K.N., 2002.Diversity and distribution of endemic and exotic earthworm in natural and regenerating ecosystems in the Central Himalaya, India.*Soil Biologoy Biochemistry* 32: 2045-2054.

Bhadauria,T., Ramakrishnan, P.S., 1991. Population dynamics of earthworm and their activities in forest ecosystem of north-east India.*Tropical Ecology* 7:305-318.

Bhadauria,T., Ramakrishnan,P.S., 1989. Earthwrom population dynamics and contribution to nutrient cycling during cropping and fallow phases of shifting agriculture in north east.*Indian Journal of Applied Ecology* 26: 505-520.

Brussaard, L., Behan. Pelletier, V.M., Bignell, D.E., 1997.Biodiversity and ecosystem functioning in Soil.*Ambio.* 26(8):563-570.

Chaudhuri, P.S., Bhattacharjee,G., 2005.Earthworms of Tripura (India).*Ecology Environment and Conservation* 11:295-301.

Chaudhuri, P.S., Nath, S., Paliwal, R., 2008. Earthworm population of rubber plantations (*Heveabrasiliensis*) in Tripura, India. *Tropical Ecology* 49(2):225-234.

Chaudhuri, P.S., Pal, T.K., Bhattacharjee, G., Dey, S.K., 2003. Rubber leaf litters (*Heveabrasiliensis*) as vermiculture substrate for epigeic earthworm *Perionyxexcavatus*, *Eudriluseugenie* and *Eiseniafoetida. Pedobiologia* 47:796-800.

Cousions, S. H.,1991.Species diversity measurement choosing the right index.*Trends in Ecology and Evolution* 6:190-192

Dash, M.C., Patra, V.C., 1977. Density, biomass and energy budget of a tropical earthworm population from a grassland site in Orissa, India. *Revus d'' EcologeetBiolgic du sols'* 14:1461-471.

Dash, M.C., Senapati, B.K., 1985.Vermitechnology Potentiality of Indian earthworms for Vermicomposting and Vermifeed. In: Mishra, K and Kapoor, S.S. (eds.) Proceedings of National Seminar on Current Trends in Soil Biology. Hissar Agriculture University, Hissar, India 61-69.

Dash, M.C.,1999. Earthworm Diversity and Ecophysiology to Vermitechnology for Sustainable development, Presidential address, Indian Science Congress, 86thSession,India.

Edwards, C.A., Bohlen, P.J., 1996. Biology and Ecology of Earthworm. Chapman and Hall, London.

Fragoso, C., Lavelle, P., 1992. Earthworm communities of tropical rain forest.*Soil Biology and Biochemistry* 24:1397-1408.

Fragoso, C., Lavelle, P., Blanchart, E., Senapati, B.K., Jimeniz, J.J., Martinx, M.A., Decaens, T., Tondon, J., 1999. Earthworm communities of tropical agro ecosystem: origin, structure and influence of management practices. In: Laveele, P., Brussard, L., Hendrix. P., (eds.) Earthworm Management in Tropical Agro-ecocsystems.CAB International, Wallingford, U.K.,pp 27-55.

Fragoso,C.,Brown,G., Patron,G.G., Blanchart, J.C., Lavelle, E., Pashanasi, P., Senapati, B., Kumar, T., 1997.Agriculture intensification, soil biodiversity and agro-ecosystem function in the tropics: The role of earthworms. *Applied Soil Ecology* 6:17-35

Gates, G.E., 1972. Burmese earthworms: An introduction to the systematics and biology of megadrileOligochaetes with special reference to southeast Asia. *Transactions of American Philosophical Society*62:1-326.

Gonzaliz, G., Garcia, E., Cruz, V., Boges, S., Zalamea, M., Rivera, M.M., 2007.Earthworm Communities along an elevation gradient in northeastern Puerto Rico.*European Journal of Soil Biology* 43:24-32.

Jameeson,B.G.M.,1978. Phylogenetic and phonetica Systematic of the OpisthoporousOligochaeta (Annelid: Clitellata). *Evolution Theory* 3:195-235.

Julka, J.M., 1976. Studies on the earthworms collected during the Dahabum Expedition in Arunachal Pradesh, India. *Records of Zoological Survey of India* 69:229-239.

Julka, J.M., 1981. Taxonomic studies on the earthworms collected during the subansiri expedition in Arunachal Pradesh, India. *Records of Zoological Survey of India* 28:1-37.

Julka, J.M., 1988. The fauna of India and the adjacent countries Megadrile, Oligochaeta (Earthworms),Haplotaxida:Lumbricina.Megascolecoidea:Octochaetidae.*Records of Zoological Survey of India*14:1-400.

Julka, J.M., Halder, K.R., 1975. On the occurrence of giant earthworm *Perionyxmacentoshi*Beddard (Oligochaeta: Megascolecidae) in Bhutan and Arunachal Pradesh (India). *Proceedings of the Zoological Society of Calcutta* 28:15-17.

Julka, J.M.,1988. On a new genus and three new species of earthworms (Megascolicidae: Oligochaeta) from Northeast India. *Research Bull. (Sci.) Punjab Univ.* 39(1-2):85-92.

Julka,J.M., Paliwal,R., 2005. Distribution of earthworms in different agro-climatic regions of India. In: Ramakrishna, G., et al. (eds.) Soil Biodiversity, Ecological processes landscape. Oxford, ABH Publications Co. Pvt. Ltd., New Delhi, India.

Lavelle, P., 1974. LesversdeterredelaSavanne de lamto in analyse dun Ecosystem Tropical Humide: la savanne de lamto (Cote d' Lvoire). *Bull. De Liaison des chercheurs de lamto*5:133-136.

Lavelle, P., Spain, A.V., 2001. Soil ecology, Kuewer Academic Publishers, Dardrecht, The Netherlands.

Lee, K.E., 1985. Earthworms: Their Ecology and Relationship with Soil and Land use. Academic Press, Sydney.

Mishra, K.C., Ramakrishnan, P.S., 1988. Earthworm population dynamics in different Jhum fallows developed after slash and burn agriculture in north-eastern India. *Proceedings Indian Academy of Sciences* 97: 309—318.

Mishra,P.C.,Dash,M.C.,1984.Population dynamics and respiratory metabolism of earthworm population in subtropical woodland of Western Orissa, India. *Tropical Ecology* 25: 103-116.

Patiri, B., Barah, A., 2007. Wild edible plants of Assam, Director Forest Communication, Forest Department, Assam, India, pp 1-463.

Reynolds, J.W., Cook, D.G., 1976. NomenclaturaOligochaetologica: A Catalogue of the names, descriptions and type specimens of the Oligochaeta. The University of New Brunswick, Fredericton, Canada, pp 217.

Reynolds, J.W., Julka, J.M., Khan, M.N., 1995. Additional earthworm records from Bangladesh (Oligochaeta: Glossoscolecidae, Megascolecidae, Moniligastridae, Ocnerodrilidae, Octochaetidae) *Megadrilogica* 6:51-62.

Sathianarayanan, A., Khan, A.K., 2006. Diversity distribution and abundance of earthworms in Pondicherry region.*Tropical Ecology* 47:139-144.

Senapati, B. K., 1993. Vermitechnology in integrated waste biomass management: Concepts Prospects and Problems. In: Biogas Slurry Utilisation Consortium on Rural Technology. Institute of Information and Technology, New Delhi, pp 57-74.

Senapati,B.K.,Dash,M.C.,1981. Effect of grazing on the elements of production in vegetation and Oligochaeta of a Tropical Pasture.*Review Ecological Biological Soils* 18:487-505.

Sinha, B., Bhadauria, T., Ramakrishnan, P.S., Saxena, K.G., Maikhuri, R.K., 2003.Impact of landscape modification on earthworm diversity and abundance in the Hariyali scared landscape, Garhwal Himalaya.*Pedobiologia* 47: 357-370.

Stephenson, J., 1923. Oligochaeta. The Fauna of British India including Ceylon and Burma. Taylor and Francis Ltd, London, pp 518.

Verma, D., Bharti, S., Yadav, S., 2010.Earthworm Resources in the Gangetic plains of Uttar Pradesh,India. *International Journal of Biodiversity and Conservation* 2 (6): 134-139.

Verma, D., Shweta, 2011. Earthworm Resources of Western Himalayan region, India.*InternationalJ. Soil Science* 6:124 – 133.

Yumnam, J.Y., 2008. Rich biodiversity of northeast India needs conservation. *Current Science* 95(3):397.

In: Native Species
Editors: L. Marin and D. Kovač

ISBN: 978-1-61470-613-7

Chapter 6

ATRIPLEX HALIMUS SUBSP. *SCHWEINFURTHII* (CHENOPODIACEAE): A NATIVE SPECIES IN SALT STEPPES OF ALGERIA – A REVIEW

Nedjimi Bouzid*
Université de Djelfa, Faculté des Sciences de la Nature et de la Vie,
Cité Aîn Chih, Djelfa, Algérie

ABSTRACT

Atriplex halimus subsp. *schweinfurthii* (Chenopodiaceae) is perennial spontaneous species of Algerian salt steppes with an excellent tolerance to drought and salinity. This species is very palatable and is a preferred fodder shrub of livestock during dearth periods. Endowed with a complex root system and a considerable biomass, *A. halimus* is an efficient and a relatively cheap barrier against erosion and desertification. Its ligneous wood is interesting source of energy. In view of this, *A. halimus* plays a major role in the rehabilitation of the degraded lands, especially in the Mediterranean arid and semi-arid areas. Salt-affected soils will remain unproductive unless efforts are made to rehabilitate them with highly salt-tolerant plants (halophytes). It goes without saying that salty areas such as

* Dr. Nedjimi Bouzid, Université de Djelfa, Faculté des Sciences de la Nature et de la Vie, Cité Aîn Chih, BP 3117 Djelfa 17000, Algérie. Tel.: + 213 662 128 131. Fax: + 213 27 900 201. E-mail address: bnedjimi@yahoo.fr.

salt steppes of Algeria can benefit from being re-vegetated by planting saltbush and farmers can also obtain fodder for their livestock from this halophytic shrub. This paper is to review the importance of *A. halimus* as forage crop for possible cultivation in saline steppes of Algeria.

Keywords: *Atriplex halimus,* Algeria, Arid salt steppes, Desertification, Halophytes, Livestock

ABBREVIATIONS

ANA: acid naphthalene acetic,
ADF: acid detergent fiber,
ADL: acid detergent lignin,
CF: crude fiber,
CP: crude protein content,
DM: dry matter,
DMD: dry matter digestibility,
NDF: neutral detergent fiber,
NFE: nitrogen free extract.

1. INTRODUCTION

Livestock production practice in Algerian steppes is mainly base on grazing natural vegetation. The native rangelands of the country are greatly affected by annual rainfall precipitation which is irregular and poorly distributed. The rangelands are characterized by a short rainy season usually not more than three to four months per year. The palatable and good quality forage always deteriorates or disappears as a result of overgrazing. Therefore, forage scarcity is prevalent and there is an urgent need for increase in feed resources in the arid zones of the country.

Native rangelands of Algerian steppes are degraded through activities associated with inadequate agriculture and overgrazing. Rehabilitation of the degraded land is generally limited by low rainfall, unfavorable soil physical conditions and often by salinity (Nedjimi, 2011).

The shortage of animal feeds is the main constraint to increase indigenous animal production, particularly in arid and semi-arid regions. Animal production, as the main source of income for nomads, is based mostly on the natural vegetation for feeding sheep, goats and camels. The palatable plant species are always overgrazed due to the grazing pressure of animals. Halophytic plants such as *Atriplex spp.*, *Suaeda spp.* and *Salsola spp.*, are generally the most palatable chenopod shrubs in several salt marshes (El Shaer 2009). Such plants are extremely valuable as a fodder reserve particularly during drought season.

Halophytes plants may provide sensible alternatives for many developing countries (Khan and Duke, 2001). These plants can grow in saline to extremely saline habitats and have particular characteristics which enable them to evade and/or tolerate salinity by various eco-physiological mechanisms. These plants are naturally grown or cultivated in salt affected lands. The vegetative yields of halophytes species could have great potentialities particularly as sources of livestock fodders. However, the value of certain halophytes has been recognized by their incorporation in the rangelands improvement programs inmany salt-affected regions throughout the world. Although economic consideration of halophytes plants is just beginning, they are now receiving increased attention particularly in arid regions where salinity problems are very crucial. There have been recent advances in selecting species with high biomass and protein levels and the ability to survive a wide range of environmental conditions including salinity.

Atriplex species (saltbushes) are dominant in many arid and semi-arid regions of the world (Figure 1), particularly in habitats that combine relatively high soil salinity with aridity (Le Houérou, 2000). Most saltbush species prosper in areas with annual rainfall ranging from 200 to 400 mm. Saltbushes have been used as a resource for domestic livestock and wildlife, and for rehabilitation of degraded lands (sand dunes, saline/alkaline soils, mine waste, badlands, shallow soils, etc.). Over 400 species of *Atriplex* have been identified on all continents. The Mediterranean Basin, with 40–50 *Atriplex* species, mostly in its southern and eastern bordering areas, is a region where saltbushes have been extensively used as fodder reserves during periods of scarcity (e.g. drought and cold periods), and as a supplementary forage resource in arid and semi-arid countries. This paper reviews the potentials of *A. halimus* subsp. *schweinfurthii* as forage for arid salt steppe of Algeria with saline ground water.

2. Plant Description

Atriplex halimus subsp. *schweinfurthii* (Chenopodiaceae) is C_4 perennial shrub, dominant in many arid and semi-arid regions of the world, particularly in habitants that combine relatively high soil salinity with aridity and have been used as a resource for livestock or for the rehabilitation of degraded lands.

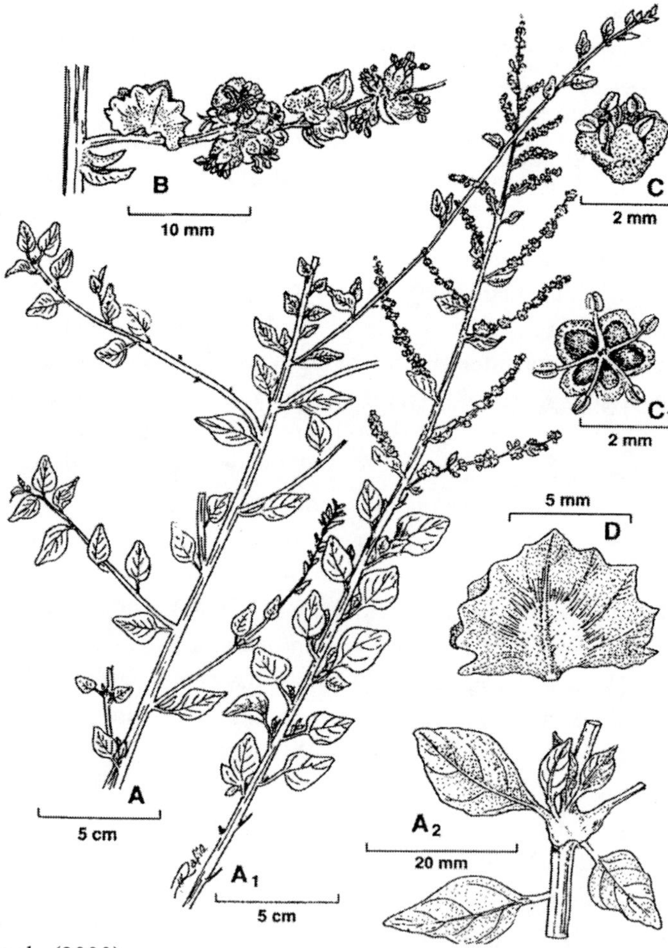

Al-Turkis et al., (2000).

Figure 1. *Atriplex halimus* subsp. *schweinfurthii* [A, A_1–Habit, A_2–Leaves; B–Inflorescence; C, C_1– male flowers; D– Fruiting bracteoles].

Figure 2. Branch with Leaves of *Atriplex halimus* subsp. *shweinfurthii.*

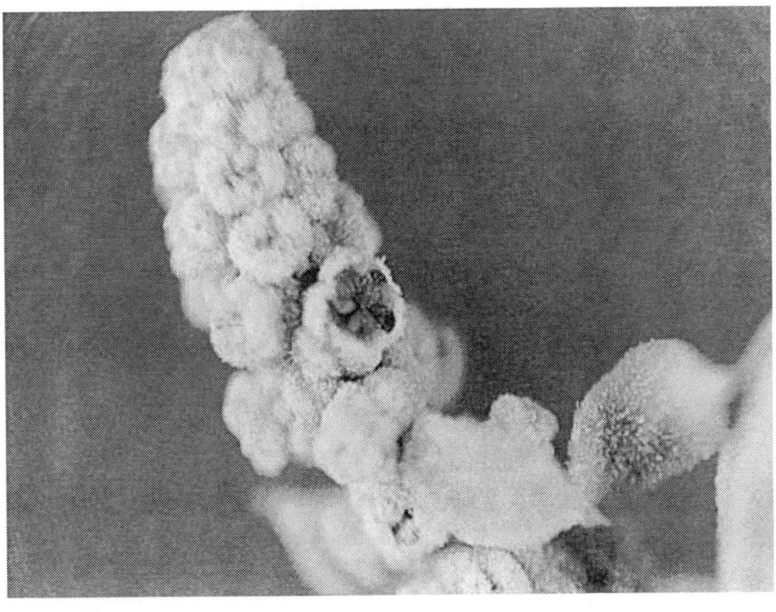

Figure 3. Inflorescence of *Atriplex halimus* subsp. *shweinfurthii.*

Erect, perennial, strong shrub, reaching up to 2.5- 03m high and spreads to 2.4m wide, with greyish white, peeling bark on the older stem. Leaves alternate, ovate-deltoid, or oblong-ovate (06 x 04cm), shiny, silver grey, entire-repand lobed, subhastate, obtuse, with a small mucro or apex emarginate, petiole (0,3-01 cm) (Figure 2). Species dioecious, that is, the sexes are borne on separate plants, inflorescence naked, usually terminal, paniculate, with many flowered congested and lateral spikes. Female flowers out numbering male flowers. Male flowers perianth small.

The flowers of saltbush are yellowish (Figure 3), inconspicuous, and unisexual, and are borne in the axils of the upper leaves or in terminal spikes. The male flowers consist of groups of stamens within a shallow 05-toothed calyx; petals are absent. Both petals and calyx are absent in the female flowers. The naked 01-seeded ovary is borne instead between 02 leaflike bracteoles.

A. halimus flower in early to late summer, and fruits ripen from early fall to winter. The flowers are wind-pollinated. The leaflike bracteoles stay green and photosynthetically active until quite late in the ripening process and probably provide resources directly to the ripening ovule within. The fruits often persist on the bushes at least until spring, and it is not uncommon to find 02 generations of fruits on a plant simultaneously. Fruit with a thick textured bract, many, condensed; bracts fused or united at base, reniform, broadly orbicular (04 x 06 mm), entire or serrate-dentate, tubercles present or absent. Seeds brown-dark brown (1.5 mm) (Al-Turkis et al., 2000).

Le Houérou (1992) divided *A. halimus* into two subspecies, *halimus* and *schweinfurthii*. This was based on differences in morphology, with *halimus* having a more erect habit, a smaller size (a height of 0.5–2.0 m, compared to 1.0–3.0 m for *schweinfurthii*), shorter fruit-bearing branches (0.2–0.5 m compared to 0.5–1.0 m) and less markedly toothed fruit valves. However, the existence of intermediate morphotypes complicates the designation of plants as one or the other subspecies. *A. halimus* subsp. *halimus* is found in semiarid and subhumid areas of the western Mediterranean and on the Atlantic coasts of France and Spain, while subsp. *schweinfurthii* occurs in arid zones with milder winters (e.g. North Africa and the eastern Mediterranean).

3. Cytotaxonomy

Ortiz-Dorda et al. (2005), studying RAPD-PCR products of DNA extracted from 51 populations of *A. halimus*, were able to separate the populations into two genetic groups which, according to morphological characters, corresponded to subsp. *halimus* and subsp.*schweinfurthii*. Based on the chromosome counts and estimates of 2C nuclear DNA for *halimus* poulations (diploid, $2n = 2x = 18$, 2C nuclear DNA = 2.40 pg) and *schweinfurthii* populations (tetraploid, $2n = 4x = 36$, 2C nuclear DNA = 5.11 pg). For certain populations having morphologies intermediate between those considered typical of these two subspecies, nuclear DNA contents showed them to be tetraploid.

4. Ecology

Atriplex halimus subsp. *schweinfurthii* is best suited to mediterranean climates, however it is a hardy species that is able to tolerate a variety of harsh environmental conditions. *A. halimus* is cold-tolerant plants, having average minimum temperatures in the coldest month of –10°C and –12°C (Aouissat et al., 2009). It has been shown recently that, under field conditions, freezing tolerance of *A. halimus* populations seemed to be related to leaf concentrations of Na and K (Walker *et al.*, 2008). Franclet and Le Houérou, (1971) indicate that *A. halimus* can survive under rainfalls of 100 - 400 mm year^{-1}, but has been reported to survive between one and several years without rainfall. Native populations are found in loam or clay depressions containing moderately saline soils and a temporary water-table, and also on outcrops of gypso-saline marls. The species grow in the soil which electric conductivity reaches 05 to 100 dS m^{-1} in surface horizons.

5. Propagation

In general, Seeds of *A. halimus* are characterized by dormancy and complex multiple dormancy mechanisms. The treatments that have sometimes been found to remove dormancy include scarification and leaching. Scarification apparently acts by weakening the bracteole walls. Actual rupture of the membranous utricle wall is usually damaging to the seed (Redondo-Gomez et al., 2007). The bracteole walls may also be weakened by steeping in sulfuric acid or by the action of saprophytic fungi under field conditions. Hand-removal of the bracteoles promotes increased germination in *A. halimus* but does not necessarily remove dormancy completely (Mâalem and

Rahmoune, 2009). The failure of excised fruits to germinate suggests a chilling requirement. Leaching probably promotes germination by removing some inhibitor from the fruit, either inorganic salts such as sodium chloride (Nedjimi et al. 2007) or an organic inhibitor sach as saponin (Nord and Van Atta, 1960). It is important to remove excess water after soaking, as germination can be inhibited by inadequate aeration.

The complex dormancy mechanisms shown by saltbush function both to time germination appropriately within a given year and to ensure carryover of a persistent seedbank between years (Ungar, 1995). Seed pretreatments to circumvent these mechanisms have limited application in field plantings but may be useful in seed quality evaluation and in nursery propagation.

Seed quality evaluation in *A. halimus* is complicated by dormancy problems. Seedlots are usually cleaned to high purity, making the purity analysis quite simple. For *A. halimus* saltbush, we proposed a 7-day germination test at 15 °C, a recommendation that was subsequently accepted as the official testing procedure (Nedjimi et al., 2006). The most important determinant of viability is the fill percentage, that is, the proportion of fruits that contain an undamaged seed with a well-developed embryo. Post-test viability determination is essential in *A. halimus*.

In practice, few seedlots of saltbush are evaluated using a germination test. The general method is to soak the intact (bracteoled) fruit for several hours or overnight and to extract the utricle by prying open the bracts, clipping at the stem end, or off-center longitudinal bisection.

Atriplex halimus can efficiently be propagated from stem cuttings especially during spring. Rooting of the new growth plant has also been reported to be more viable than older ones (Franclet and Le Houérou, 1971). Souayah et al. (2004) have shown that *A. halimus* terminal cuttings propagated and treated with ANA have quick rooting and can be transplanted into pots in nurseries. *A. nummualria* may be spaced at 02m x 03m and can be grazed by animals when it's 1.5m high i.e. second or third year of growth (Le Houérou, 1992).

Atriplex halimus may also be direct-seeded successfully (depths of 0.5 to 1 cm), the most broblemes of this methode include poor choice of ecotypes; using poor-quality seed; planting too deep; planting at the wrong season; excessive competition from weeds or seeded grasses; interactions with pathogenic fungi such as damping-off diseases; and seedling predation by grasshoppers, rabbits, or other animals. *A. halimus* fruits are apparently not particularly attractive to granivorous rodents, possibly because of the saponin content of the bracts (Meyer et al., 1998), so pre-emergence seed depredation is rarely a problem. Seeding rates of 4 to 8 kg/ha have been recommended for de-winged lots of *A. halimus*. This corresponds to about 200 to 500 seeds m^{-2}. Once seedlings establish, young plants grow rapidly and may become reproductively mature in their second growing season. Optimal season for planting varies according to precipitation patterns. In autumn and winter precipitation zones such as the Algerian salt steppe, autumn or early winter planting has been most successful (Nedjimi and Daoud, 2009).

6. Forage Production

In the Mediterranean regions, stands of *A. halimus* subsp. *schweinjxrthii* in moderately saline depressions may have a productivity of 0.5 to 05 t DM $ha^{-1}day^{-1}$ of forage (Le Houérou, 2000). The reported annual productivity of halophyte communities varied from 0.5 to 09 t DW/ha. The variability in productivity is related to the growing environment including ecological conditions of soil and climate and management applied. Palatable biomass of *A. halimus* in Algerian salt steppe averaged about 02 to 05 t DM $ha^{-1}day^{-1}$, the average precipitation in these areas is 250-300 mm $year^{-1}$ (Franclet and Le Houérou, 1971).

7. Nutrient Values

Mean levels of DM, ash, CP, CF, ADF, NDF, NFE, ADL and DMD in *A. halimus* species are shown in Table 1. *A. halimus* plants are relatively high in crude protein (CP) and ash (12.6% DM and 22.7% DM respectively) as a character for halophytic plants. Digestibility of *A. halimus* species on dry matter (DMD) may reach 56%. Based on mentioned data, it seems that *A. halimus* could cover the essential nutrients for maintenance requirements of small ruminants in good rainy seasons according to the recommended nutritional requirements of livestock. *A. halimus* plants attained a relative high content fiber constituents such as NDF, ADF, ADL, as summarized in Table 1, high levels of ADL and NDF have been considered the most remarkable factors which limit forage intake and digestibility (El Shear, 2006). It is reported that fibrous materials and ash contents of halophytic feed materials are higher and increase while gross energy and protein contents are low and decrease with advancing maturity of the plants. Therefore, most of halophytes are nutritious in wet seasons and can overcome maintenance requirements of animals whereas they are low in nutrients in summer and autumn (dry season) and need to be supplemented with other feed ingredients, particularly with energy feed resources (El Shaer, 2010).

Mean levels of sodium (Na), calcium (Ca), potassium (K), magnesium (Mg), Phosphorus (P) and Sulfure (S) in *A. halimus* are shown in Table 2. *A. halimus* plants are relatively high in most of major minerals without any risk for animal health (Gihad and El Shaer, 1994). However, this plants are deficient in sulphur and phosphorus (El-Shatnawi and Turuk, 2002).

Table 1. Nutritive values (% DM basis) and dry matter digestibility (%) of *A. halimus*

Chemical composition	DM	CP	CF	Ash	NFE	NDF	ADF	ADL	DMD
A. halimus	34.2	12.6	25.4	22.7	37.0	64.6	40.2	10.4	56.7

Data cited from El Shaer (2010).

Table 2. Mineral composition (% DM basis) of *A. halimus*

Mineral composition	Na	Ca	K	Mg	P	S
A. halimus	4.41	1.51	1.33	0.32	0.29	0.17

Data cited from Gihad and El Shaer (1994).

Most of halophytes are characterized by high ash content, mainly Na, K, Ca and Silica. As stated by Masters et al. (2001), sheep can tolerate a NaCl intake in feed of about 100–150 g day^{-1} provided that they have access to a non-saline drinking water. It means that the high level of dietary salt will act to reduce feed consumption. Sheep grazing *Atriplex ssp.* consume 07 to 7.5 liter day^{-1} of water in comparison to 3.2 liter day^{-1} on grasslands (Arieli et al., 1989). El-Shatnawi and Turuk (2002) reported that increased consumption of *A. halimus* by animals was accompanied by higher water intakes for urinary excretion of Na. The high Na and K content may limit feed intake and reduce digestibility by shortening rumen turnover times (Rossi et al., 1998). High salt content is perhaps the major negative component in *A. halimus*. Sodium levels could be higher than 4% DM. Therefore, mixing saltbushes with low salt forages is desirable (El Shear, 2010) and recommended for better utilization of *A. halimus* saltbush. Moreover, offering fresh drinking water to animals fed saltbush would reduce the stress of salt intake and also enhance saltbush consumption and nutrients utilization (Wilson, 1996).

8. Palatability and Preference

Palatability is defined as the ratio between the amounts of feed ingested by herbivores and the amount on offer, for a given period of time. Preference is the order in which forage species are selected by herbivores within a given plant community or population, or at a given site, at a given time (Le Houérou, 2000). These concepts, however, are liable to wide variability in time and space, depending, on many variables and parameters that may change with season, site, animals and other local conditions. Some of these variables were linked to the plant, others to the animal and environmental factors. For a given species palatability for a given type of cattle varies with phenological stage, the organ concerned and the season (El-Shatnawi and Turuk, 2002).

As a rule of thumb, the content of crude fibre in forage plays an important role in its selection by livestock. Forages with high fibre content are usually better accepted by cattle, also this depends on the high levels of protein in the overall diet (El Shaer, 2010). The stage of growth and maturity considerably affect the nutritive value, palatability and utilization of *A. halimus*. Such plants were nutritious in wet season while they are poor during dry season (El Shaer, 2006).

As a supplementary fodder, *A. halimus* should not take more than 25-30% of the sheep's diet. El Shaer (2010) suggested that the high salt content of saltland forage plants is likely to be the major determinant of palatability and that dilution of salt content through the availability of other feed resource would be necessary to improve intake and performance. Finishing sheeps on saltbush and either supplemented with hay or grain did not present any apparent meat quality problems compared to finishing sheeps on Lucerne (Hopkins and Nicholson, 1999). *Atriplex* contains considerable amounts of CP (12.6%) and crude fibre (25.4%) with digestibility of 56.7%. El Shear (2006) reported that sheep fed on *A. halimus* had significant weight gain in semi-arid region of Egypt, where the soil is very saline, but water should be available to prevent Na toxicity (Wilson, 1996).

Conclusion

This paper reviews the potentials of *A. halimus* subsp. *schweinfurthii* as forage crop for possible cultivation in saline steppes of Algeria. This species can produce relatively high consumable biomass in saline areas where non-halophytic species cannot grow or have low dry matter yields. Therefore, *A. halimus* can provide a drought reserve or a supplementary feed source under arid and semi-arid conditions. On grazing lands, the species can serve as a complementary nutrients source to other conventional feedstuffs, such as cereal straws or hays. This plant could be used advantageously as alternative feeds to replace totally or partially common feedstuffs, thus to alleviate feeding cost. However, the presence of high contents of ash and minerals should be taken into consideration when formulating diets containing *A. halimus* for Livestock.

Acknowledgments

Algerian Ministry of Higher Education and Scientific Research is gratefully acknowledged (Project no. F–02820100012).

References

Arieli A., Naim E., Benjamin R.W., Pasternak D. (1989) The effect of feeding saltbush and sodium chloride on energy metabolism in sheep. *Animal Production* 49: 451–457.

Aouissat M., Walker D.J., Belkhodja M., Fares S., Correal E. (2009) Freezing tolerance in Algerian populations of *Atriplex halimus* and *Atriplex canescens*. *Spanish Journal of Agricultural Research*, 7(3): 672–679.

Al-Turkis T.A., Omer S., Ghafoor A. (2000) A synopsis of the genus *Atriplex* L. (Chenopodiaceae) in Saudi Arabia. *Feddes Repertoriurn* 111: 261–293.

El Shaer H.M. (2006) Halophytes as cash crops for animal feeds in arid andsemi-arid regions. In: Öztürk M., Waisel Y., Khan M.A., G. Görk (eds) *Biosaline agriculture and salinity tolerance in plants*. Birkhäuser Verlag Publishers: Switzerland, 117–128.

El Shaer H.M. (2009) Potential Role of Sabkhas in Egypt: An Overview. In: Ashraf M., Ozturk M., Athar H.R. (eds) *Salinity and Water Stress.* Springer Publishers, 221–228.

El Shaer H.M. (2010) Halophytes and salt-tolerant plants as potential forage for ruminants in the Near East region. *Small Ruminant Research* 91: 3–12.

El-Shatnawi M.J., Turuk M. (2002) Dry matter accumulation and chemical content of saltbush (*Atriplex halimus*) grown in Mediterranean desert shrublands. *New Zealand Journal of Agricultural Research* 45: 139–144.

Franclet A., Le Houérou N.H. (1971) *Les Atriplex en Tunisie et en Afrique du Nord*. Doc. Tech n° 7. FAO. Rome, 249p.

Gihad E.A., El Shaer H.M. (1994) Nutritive value of halophytes. In: Squires V.R., Ayoub A.T. (eds) *Halophytes as a resource for livestock and for rehabilitation of degraded lands.* Kluwer Academic Publishers: The Netherlands, 281–284.

Hopkins, D. L. and A. Nicholson, 1999. Meat quality of wether lambs grazed on saltbush *(Atriplex nummularia*) plus supplements or Lucerne (*Medicago sativa*). *Meat Science* 51: 91–95.

Khan M.A., Duke N.C. (2001) Halophytes-A resource for the future. *Wetlands Ecology and Management* 6: 455–456.

Le Houérou HN. (1992) The role of saltbushes (*Atriplex* spp.) in arid land rehabilitation in the Mediterranean Basin: a review. *Agroforestry Systems* 18: 107–148.

Le Houérou H.N. (2000) Utilization of fodder trees and shrubs in the arid and semi-arid zones of west Asia and north Africa. *Arid Soil Research and Rehabilitation* 14: 101–135.

Mâalem S., Rahmoune C. (2009) Toxicity of the salt and pericarp inhibition on the germination of some *Atriplex* species. *American-Eurasian Journal of Toxicological Sciences* 1(2): 43–49.

Masters D.G., Norman H.C., Dynes R.A. (2001) Opportunity and limitations for animal production from saline land. *Asian-Australian Journal of Animal Science* 14: 199–211.

Meyer S.E., Carlson S.L., Garvin S.C. (1998) Seed germination regulation and field seed bank carryover in shadscale (*Atriplex confertifolia)*: Chenopodiaceae. *Journal of Arid Environments* 38: 255–267.

Nedjimi B., Daoud Y., Touati M. (2006) Growth, water relations, proline and ion content of in vitro cultured *Atriplex halimus* subsp. *schweinfurthii* as affected by $CaCl_2$. *Communications in Biometry and Crop Science.* 1(2): 79–89.

Nedjimi B., Daoud Y., Touati M. (2007) Effet du stress salin sur la germination, la croissance et l'accumulalion ionique chez des jeunes plants d'*Atriplex halimus* subsp. *schweinfurthii* en culture *in vitro*. *Biotechnologie Végétale* 1: 49–55.

Nedjimi B., Daoud Y. (2009) Effects of calcium chloride on growth, membrane permeability and root hydraulic conductivity in two *Atriplex* species grown at high (sodium chloride) salinity. *Journal of Plant Nutrition* 32: 1818–1830.

Nedjimi B. (2011) Rangeland improvement and management options in the arid steppes of Algeria. In: Germano M. Denise (eds) *Steppe ecosystems: Dynamics, land use and conservation*, Nova Science Publishers, Inc: New York (in press).

Nord E.C., Van Atta G.R. (1960) Saponin: a seed germination inhibitor. *Forest Science* 6: 350–353.

Ortiz-Dorda J., Correal E., Cenis J.L. (2005) Genetic variation of *Atriplex halimus* in the Mediterranean basin. *Annals of Botany* 95: 827–834.

Redondo-Gomez S., Mateos-Naranjo E., Wharmby C., Luque C.J., Castillo J.M., Luque T., Mohamed M.F., Davy A.J., Figueroa M.E. (2007) Bracteoles affect germination and seedling establishment in a Mediterranean population of *Atriplex portulacoides*. *Aquatic Botany* 86: 93–96.

Rossi R., Del Prete E., Rokitzky J., Scharrer E. (1998) Effect of high NaCl diet on eating and drinking patterns in Pygmy goats. *Physiology Behavior* 63: 601–604.

Souayah N., Khouja M.L., Rejeb M.N., Bouzid S. (2004) Micropropagation d'un arbuste sylvo-pastoral, *Atriplex halimus* L. (Chénopodiacées). *Cahiers Options Méditerranéennes*. 62 : 131–135.

Ungar, I.A. (1995) Seed germination and seed bank ecology of halophytes. In : Kigel J., Galili G. (eds) *Seed development and germination*. Marcel Dekker: New York, 599–627.

Walker D.J., Romero P., de Hoyos A., Correal E. (2008) Seasonal changes in cold tolerance, water relations and accumulation of cations and compatible solutes in *Atriplex halimus* L. *Environmental and Experimental Botany* 64: 217–224.

Wilson, A. D. (1996) The intake and excretion of sodium by sheep fed on species of *Atriplex* (saltbush) and *Kochia* (Bluebush). *Australian Journal of Agricultural Research* 17 : 155–163.

INDEX

A

B

C

D

E

F

G

H

I

O

P

E

F

G

H

I

T

U

V

W

Z

Q

R

S